D1618632

BASIC PRINCIPLES OF WATER TREATMENT

C. D. Morelli

TALL OAKS PUBLISHING INC.
P.O. Box 621669
Littleton, Colorado 80162, U.S.A.

Library of Congress Catalog Card Number: 95-060295

ISBN 0-927188-5-8

FIRST EDITION

TALL OAKS PUBLISHING, INC.
P.O. Box 621669
Littleton, Colorado 80162
U.S.A.
303/973-6700

PRINTED IN THE UNITED STATES OF AMERICA
Front cover: Pretreatment filtration equipment; Courtesy: United States Filter Corporation.

This book is dedicated to the memory of my wife, Leah.

PREFACE

It would surely be the height of arrogance for this author to claim an all-knowing role as an expert in all of the subjects covered in this text.

Nevertheless, the material contained herein reflects a wide ranging experience of more than fifty years of close association with the water and waste problems of the manufacturing industries.

Through these years, colleagues, clients and equipment manufacturers contributed to a continuing practical education on these subjects.

To them and to the companies who have graciously supplied illustrations and information for this volume, I am truly grateful.

My deep thanks to Judy Smith, whose close attention to details has kept this work "on track".

Cliff Morelli

ACKNOWLEDGEMENTS

A special word of gratitude goes to Jerry Keller, President and Publisher of Beverage World, Great Neck, New York, for his generous permission to use certain materials from the second edition of th Water Manual which was authored by me. I would also like to acknowledge the following for providing material for use in this book.

American Water Works Association, Denver, Colorado
Aquamatic Inc., Rockford, Illinois
Chemical Week magazine, New York, New York
Cla-val Co., Newport Beach, California
Coopermatics Inc., Northampton, Pennsylvania
Filtomat, Orival Inc., Englewood, New Jersey
Goulds Pumps, Seneca Falls, New York
Hungerford & Terry, Clayton, New Jersey
Ideal Horizons, Rutland, Vermont
Infilco Degremont, Richmond, Virginia
International Society of Beverage Technologists, Hartfield, Virginia
Ionics Inc., Watertown, Massachusetts
Nalco Chemical Co., Naperville, Illinois
NASQUAN (U.S. Dept. of Interior), Washington, D.C.
National Lime Association, Arlington, Virginia
Nickerson & Collins, Chicago, Illinois
Osmonics Inc., Minnetonka, Minnesota
Ozonia North America, Richmond, Virginia
U.S. Department of Agriculture, Washington, D.C.
U.S. Environmental Protection Agency, Washington, D.C.
U.S. Filter Co., Rockford, Illinois
U.S. Geological Service, Washington, D.C.
Wallace & Tiernan, Belleville, New Jersey
Westvaco Corp., Covington, Virginia

CONTENTS

CHAPTER 1

WATER SOURCES

Rain and snow are sources of our natural waters. As they fall from the clouds to the earth, they are already absorbing or picking up the carbon dioxide, impurities, and organisms that abound in the atmosphere. Figure 1-1 shows the natural cycle of evaporation and condensation of water. This representation is a simplification, however, which could lead one to believe that only pure water, containing no dissolved substances, is involved in the process. To the contrary, rain and snow are becoming polluted the instant they are formed.

Table 1-1 shows an interesting analysis of rain, hail, and snow. Neither Figure 1-1 nor Table 1-1 indicate the presence of any organic substances in runoff water,

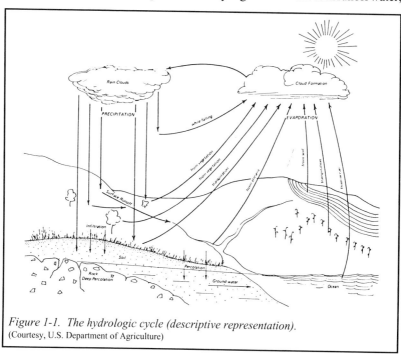

Figure 1-1. The hydrologic cycle (descriptive representation).
(Courtesy, U.S. Department of Agriculture)

whether it culminates in lakes, in reservoirs, or as well water. It is known that most of our sources of water, whether surface or groundwater, do contain organics of one type or another.

Some fairly complicated interactions take place during the hydrologic cycle. *Evaporation, condensation, precipitation, transpiration, runoff, soil water, outflow,* and *subsurface outflow* are familiar terms to anyone who works with water supplies, and are explained and interrelated in Figure 1-2.

While this water in the hydrologic cycle has been collecting as surface supplies or while it is circulating in the ground to form subsurface supplies (wells), it becomes polluted with organisms, organic compounds, and inorganic compounds. It becomes evident that water from rain and snow, as it collects in rivers, lakes, reservoirs, and wells, cannot always be used as is for human consumption; nor, without proper preparation, can it be used in the manufacture of many products.

The various sources of water supplies will be discussed separately so that the reader will be made aware of their differences and similarities. Before proceeding, however, a caveat: It is not possible to set specific values for the chemical characteristics of each of the sources. For example, each source, depending upon its location, will contain varying quantities of characteristics such as hardness and alkalinity.

It seems that the most consistent source for human and product use is the water

Table 1-1
Analysis of Rain, Snow, and Hail

(All results in ppm)	Rain After 4-hr Continuous Rain	Rain After 22-hr Continuous Rain	Snow	Hail
Total hardness as $CaCO_3$	43	8	18	28
Calcium hardness as $CaCO_3$	42	8	14	25
Magnesium hardness as $CaCO_3$	1	—	4	3
Alkalinity as $CaCO_3$	19	5	—	4
Sodium as Na	5	0.11	5	—
Ammonia as NH_3	1.5	2	6	1
Chlorides as Cl	7	4	12	7
Sulfates as SO_4	26	3	21	17
Nitrates as NO_3	1	—	1	—
Iron as Fe	0.9	0.1	1.2	2.4
Silica as SiO_3	0.15	0.15	3	1

$CaCO_3$ - calcium carbonate
(Courtesy, U.S. Filter/Permutit)

that municipalities and private purveyors deliver to their consumers; however, these supplies also have widely ranging chemical characteristics. Tucson, Arizona, delivers water to its customers with a hardness of 141 parts per million (ppm) and an alkalinity of 150 ppm. New York City, on the other hand, delivers water with hardness and alkalinity of approximately 30 ppm (all values as $CaCO_3$, a comparison that will be explained in Chapter 3, "Units of Measurements and Useful Calculations").

Surface Supplies

Rivers. Some of the water reaching earth will go directly into the ground, some will be immediately impounded in lakes and reservoirs, and some will collect as runoff to form streams and rivers that will then flow to the oceans. Throughout

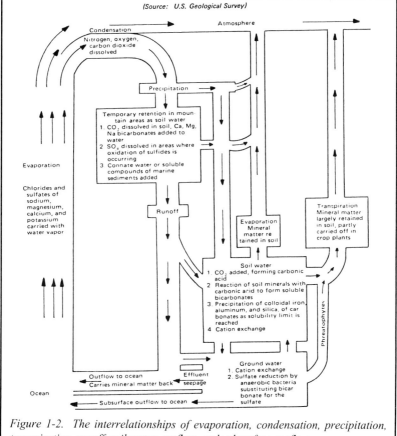

Figure 1-2. The interrelationships of evaporation, condensation, precipitation, transpiration, runoff, soil water, outflow, and subsurface outflow.
(Courtesy, U.S. Geological Survey)

this process, the water is constantly dissolving different types and various quantities of those substances that make up the earth's crust (inorganic) and its vegetation (organic).

Included among these substances could be pollutants placed there by our civilization These man-made pollutants will be discussed separately.

The National Stream Quality Accounting Network (NASQAN), a division of the Department of the Interior, classifies the *inorganic* substances as "common constituents." These substances generally dominate the total mass of dissolved

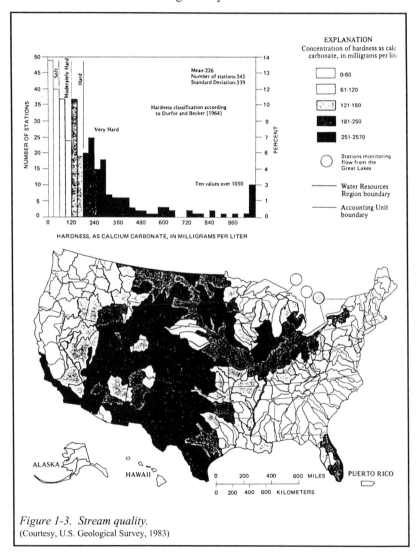

Figure 1-3. Stream quality.
(Courtesy, U.S. Geological Survey, 1983)

material.

Those *organics* that are of concern are basically bacteria, algae, yeasts, color, and humic and fulvic acids. All of these are harmful in almost every manufacturing process.

The surveys by NASQAN "summarize the water qualities of the rivers of the United States, and these data represent the quality of the water at the particular sampling points at that particular time."

The maps and graphs in Figure 1-3 indicate the concentrations of hardness, alkalinity, chlorides, and sulfates in samples collected from river locations throughout the United States.

These illustrations give an idea of what happens to rain and snow collected in rivers over the different areas of the country. They indicate that, depending upon the composition of the terrain over which identically composed rain or snow take place, there are extreme variations in the composition of the water that becomes a river's supply.

Figure 1-3 is a representation of a general nature, used here only to indicate the concentrations of various components. It is not possible to obtain specific data for a specific location unless water is impounded at a sampling point. The information shown is used only to give the reader some idea of the quality of surface water runoff throughout the United States. (*Runoff* is defined as the area over which the water for impoundment is collected.)

Rivers are extremely important sources of water, and utilities in many heavily populated areas find it necessary to go long distances to reach rivers that will satisfy their needs for water. Large sections of Arizona and California use water taken from the Colorado River, hundreds of miles away. Likewise, New York City, located on the brackish Hudson River, gets its water in part from the more distant Delaware River. New York City zealously and expertly oversees both the source and the runoff that form its reservoirs, to the extent that it has controlled (and offered every available aid and cooperation) to the sewerage treatment plants that spill their effluents into the New York City system.

Lakes and reservoirs. Included with lakes and reservoirs as containments are dams and ponds (not really significant), and all have some characteristics similar to those of a river water. The quality of water in a lake at any point along a river would be basically the same as the water that would have flowed into the river at that particular location. The data used for the NASQAN illustrations (Figure 1-3) include that from containments such as lakes.

Although the various types of bodies of containment are similar, they have some differences:

● Lakes are usually a product of nature, while reservoirs and most dams are man-made.

● Lakes as a rule are larger than reservoirs and dams, although some reservoirs may be larger than many lakes.

● Lakes generally contain waters that are the result of runoff, which may or may not deliver a suitable water.

● Lakes supplies are most likely to be used in those areas in which underground or river supplies are unavailable, do not have sufficient capacity, or are of such chemical quality to make them undesirable.

It would be gratifying to be able to make a definitive statement on the quality of the water contained in these impoundments. However, a search of much of the literature on this subject indicates that any conclusions are of a broad nature because of the number of variations encountered. (Average well analyses, as compared to river analyses, are discussed under "Groundwater Supply," in the section that follows.)

Chemical analyses could be exhibited here, but they would describe only individual cases that are not indicative of all surface supplies. The composition of lake water changes seasonally and sometimes even daily with weather conditions. One factor that creates change is the seasonal turnover that occurs in most lakes in the United States during the spring and fall. (Seasonal turnover is discussed in Chapter 11, "Organic Matter.")

Although the major dissolved mineral constituents may not be greatly affected by seasons and weather, such factors as dissolved oxygen, algae, temperature, suspended solids, turbidity, and carbon dioxide will change because of biological activity.

To summarize this discussion, the following statement is taken from "Water Supply and Treatment," published by the National Lime Association, Washington, D.C. (1943). Although it was made 50 years ago and is rather general, it substantially states the nature of surface supplies.

> Surface water, whether from streams, lakes or reservoirs, usually is contaminated and therefore unsafe and unsatisfactory for human consumption until properly treated. Municipalities sometimes discharge sewage into a water course that is used as a public water supply. This is perhaps the most dangerous source of contamination. Soil washings may carry mud, leaves, decayed vegetation and human and animal refuse into the supply, thus rendering it turbid or unclean in appearance. The turbidity, or muddiness and mineral content of water in flowing streams vary from day to day. Following heavy rains or freshets it may be extremely muddy and low in mineral content, whereas during the dry seasons it may be relatively clear and more highly mineralized. Surface supplies may be muddy or clear, soft or hard, depending on the season. The immensity of the problem of mud removal from water taken from surface sources may be realized when we find that many of the large communities remove from 100 to 1000 or

more tons of mud from a single day's supply of water before pumping it to the consumer. Human feces and urine may cause typhoid fever and dysentery. Organic wastes furnish food for microorganisms which include vegetable growths such as algae and the lower forms of animal life. They may impart to the water disagreeable tastes and odors.

The waste liquors from the manufacturing plants, mines and quarries are often discharged into streams. These wastes may be objectionable because they are acid in nature, in which case they render the water either unfit for use or too corrosive for distribution through ordinary iron pipe distributing systems. Industrial wastes also may contain excessive quantities of organic material, which after decomposition causes the water to be unpalatable. Substances such as phenols from coke-oven plants are sometimes discharged into streams and lakes and are especially objectionable if chlorine is used in the treatment of the water supply. Chlorophenols are produced which impart to the water a disagreeable medicinal taste.

The watersheds of many streams are underlain with limestone and other rock formations containing soluble minerals. Water flowing through these deposits may dissolve compounds, causing the water to be highly mineralized and hard.

Although the above might seem to pertain principally to flowing streams, it must be realized that in most cases streams do flow into various impoundments. As a result, constant monitoring and vigilance must be made a part of the daily operation of water treatment facilities, to assure that polluting discharges to water sources are maintained under control and within those limits presently

Figure 1-4. Groundwater: principal aquifers of the United States.
(Courtesy, U.S. Geological Survey)

mandated by the regulatory bodies that have come into force since that statement was written.

Groundwater Supplies (Well Water)

Groundwater is usually considered as being well water, although it may include water from sources such as springs and infiltration galleries. Approximately one-third to one-half of snow and rainfall will be deposited underground in bodies of water called *aquifers* (Figure 1-4), and by far the greatest portion of our freshwater supply is stored in this matter. Aquifers can be considered as underground rivers and lakes, and the wells used to draw this water to the surface number in the millions.

It has been estimated that well over one-half of the fresh water consumed in the United States emanates from these aquifers. This is a large percentage, but if one considers that a huge amount of this water is used for irrigation, this statement is not implausible.

Note in Figure 1-4 that groundwaters are plentiful in the eastern and central states; and that the aquifers in the western states are scattered and smaller. Whether or not such supplies will be used depends upon the ability and cost to bring such waters to the surface. Keep in mind, however, that approximately three-quarters of the municipalities in the United States depend upon well water. It is estimated that over three-quarters of the world's freshwater supply is contained in such aquifers.

On its way to becoming an aquifer, rainfall has long and intimate contact with the many minerals found in the stratum above that aquifer. Just as water for coffee picks up the odor, color, and flavor of coffee grounds as it percolates, rainwater will absorb characteristics of soluble substances as it trickles downward.

The temperature of groundwater coming from a depth of 50 feet is the same as the average temperature of the region under which it lies. Water from a depth of less than 50 feet will be a little colder in the winter than that average, and a little warmer in the summer. Water from a depth greater than 50 feet has a temperature higher than the average temperature of the region under which it lies. The temperature increases on an average of one degree Fahrenheit for each 60 feet of depth below 50 feet.

Individual wells, even in close proximity, can and do exhibit differences in their chemical characteristics; therefore, no specific values will be shown here, but general comparisons with surface supplies will be discussed.

Compared to surface water, groundwater supplies have the following advantages:

● They are usually clearer, with less color and suspended solids.

● They contain less bacteria and other organics.

● If taken from a single well, they usually have a uniform mineral content.

Following are some of the disadvantages of groundwater supplies:

● Their scarcity for large consumers of large amounts.

● Their uncertain qualities. Calcium and magnesium compounds are present in larger quantities than in surface waters found in the same localities. The same is probably true for other dissolved salts such as chlorides and sulfates. As a rule, the total dissolved solids (TDS) will be greater.

● Iron and manganese are present in many well supplies.

● Hydrogen sulfide is often present.

● The cost of pumping well water is usually greater than the cost of pumping surface water.

● The mineral content from two wells may be entirely different even though they are located in the same plot of ground (Table 1-2).

This mineral content difference is illustrated by wells 1, 2, and 3 in Table 1-2.

Samples 1, 2, and 3 were taken from wells on Long Island, New York, while sample 4 is of New York City water. Although wells 1, 2, and 3 are located in close proximity, wells 1 and 2 are quite similar, but both are quite different from well 3. Each of these sources delivers an excellent water (not unusual for East Coast groundwater supplies). Typical analyses of European waters are shown in Table 1-3.

Although the specific value of each chemical characteristic would be different, these European analyses are similar to descriptions of United States waters. They show well water as cooler and containing less color and dissolved oxygen than river water, but containing more hardness, alkalinity, sulfates, iron, and chlorides.

Table 1-4 summarizes the main differences between surface and groundwaters. To the summary in Table 1-4 might be added the fact that some well water can be contaminated by bacteria and other organics if the well area is insufficiently protected. Well water is usually very clear, but this is no assurance that these problems cannot exist.

Public Supplies
Municipalities and purveyors. Large municipalities make use of surface sources to secure an adequate supply of water, whereas smaller communities, private estates, and industrial plants ordinarily use underground sources. It is

almost impossible to find a source of either surface or underground water that will meet the modern requirements for a public water supply without some form of treatment.

The requirements for a public water supply are as follows:

● That it shall contain no organisms that cause disease.

● That it be sparkling clear and colorless.

● That it be good-tasting, free from odors, and preferably cool.

● That it be reasonably soft.

Table 1-2
Well Water Analyses, a Municipal Supply

Well	1	2	3	4
Turbidity (NTU)	0.17	0.32	0.18	
pH	7.3	7.2	7.3	7.5
Iron (mg/L Fe)	0.12	0.52	0.12	0.010
Manganese (mg/L Mn)	<0.01	0.02	0.12	0.200
Chloride (mg/L Cl)	7.5	6.0	17.0	20.8
Total hardness (mg/L $CaCO_3$)	33	26	102	40.0
Total alkalinity (mg/L $CaCO_3$)	28	26	80	20.0
Phenol alkalinity (mg/L $CaCO_3$)	0.0	0.0	0.0	0.0
Carbon dioxide (mg/L CO_2)	2.2	2.6	6.4	
Ammonia nitrogen (mg/L NH_3-N)	<0.01	<0.01	<0.01	
Nitrite nitrogen (mg/L NO_2-N)	<0.01	<0.01	<0.01	
Nitrate nitrogen (mg/L NO_3-n)	0.96	0.66	5.59	
Surfactants (mg/L MBAS)	<0.02	<0.02	<0.02	
Sodium (mg/L Na)	5.6	4.6	14.1	12.0
Potassium (mg/L K)	0.72	0.60	1.56	5.0
Calcium (mg/L Ca)	9.0	9.8	31.5	17.6
Magnesium (mg/L Mg)	1.95	1.64	4.70	
Specific conductance (mg/L μmho/cm)	99	91	280	
Sulfate (mg/L SO_4)	5.2	5.7	6.0	20.1
Copper (mg/L Cu)	0.03	0.08	0.54	0.05
Zinc (mg/L Zn)	<0.01	0.38	0.08	
Chromium (mg/L Cr)	<0.005	<0.005	<0.005	
Fluoride (mg/L F)	<0.1	<0.1	0.1	
Phosphate (mg/L P)	0.28	<0.1	<0.1	
Orthophosphate (mg/L PO_4)	0.19			
Metaphosphate (mg/L PO_4)	0.09			

NTU - nephelometric turbidity units
mg/L - milligrams per liter
MBAS -
(Courtesy, Long Island, N.Y. municipal supply)

- That it be neither scale-forming nor corrosive.
- That it be free from objectionable gas such as hydrogen sulfide, and objectionable minerals such as iron and manganese.
- That it be plentiful, and low in cost.
- That it contain no chemicals that could prove to be toxic or carcinogenic to humans.

This definition certainly sets forth the type of water most desired for potable

Table 1-3
Typical Analyses (European Water)

	River Water	Groundwater
Temperature (°C)	14	9.5
Turbidity (NTU)	18	0.2
Color (mg/L Pt-Co)	30	10
Suspended solids (mg/L)	25	0.35
pH	8	6.7
TAC (French degrees)	20	35
(meq/L)	4	7
TH (French degrees)	22	80
(meq/L)	4.4	16
Calcium (French degrees)	17	71
Magnesium (French degrees)	5	9
Chlorides (mg/L Cl)	25	70
Sulfates (mg/L SO_4)	18	330
Iron (mg/L Fe)	1.4	3.5
Manganese (mg/L Mn)	Trace	1.2
Ammonia (mg/L NH_4)	0.7	1.5
Nitrites (mg/L NO_2)	0.2	Trace
Nitrates (mg/L NO_2)	3	1
Free CO_2 (mg/L)	4	135
Dissolved oxygen (mg/L)	9.5	None
Oxidizability in permanganate, heated and in acid medium (mg/L O_2)	7.5	1.5

Pt-Co -
TAC -
meq/L - milliequivalents per liter
TH -
1.0 French degree = 10.0 ppm
(See Tables 3-1 and 3-2 for conversion to other units.)

use, but it places no numerical values or limits by which it can be judged. The criteria for drinking water that are accepted throughout the world are those established by the World Health Organization (WHO) in its "International Standards for Drinking Water." There are some small differences in various countries, but these are minor.

In the United States, the governing concentrations are as determined by the United States Environmental Protection Agency (EPA), and Table 1-5 shows the drinking water regulations set forth by the EPA (1990).

These parameters have changed little since those issued in 1962 by the U.S. Public Health Service (forerunner of EPA). Since that time, the only major additions have been sodium and trihalomethane.

Table 1-4
Summary of Surface and Groundwaters

Characteristics	Surface Water	Underground Water
Temperature	Varies with season	Relatively constant
Turbidity, suspended solids	Level variable, sometimes high	Low or nil
Mineral content	Varies with soil, rainfall, effluents	Largely constant, generally appreciably higher than in surface water from the same area
Divalent iron and manganese (in solution)	Usually none, except at the bottom of lakes or ponds in the process of eutrophication	Usually present
Aggressive carbon dioxide	Usually none	Often present in large quantities
Dissolved oxygen	Often near saturation level	Usually none at all
Ammonia	Found only in polluted water	Often found, without systematically indicating pollution
Hydrogen sulfide	None	Often present
Silica	Moderate proportions	Level often high
Nitrates	Level generally low	Level sometimes high; risk of methemoglobinemia
Living organisms	Bacteria (some pathogenic), viruses, plankton	Ferrobacteria frequently found

Table 1-5
Summary of National Primary Drinking Water Regulations
(as of July 1990)

Contaminant (In mg/L unless otherwise noted)	CLGs	SMCLs
Microbiological Contaminants		
Coliforms (total)	0	1/100 mL
Giardia lamblia	0	TT3
HPC	—	TT3
Legionella	0	TT3
Virus	0	TT3
Turbidity	—	1.5 NTU
Inorganic Contaminants		
Arsenic	—	0.05
Barium	—	1
Cadmium	—	0.010
Chromium	—	0.05
Fluoride	4.0	4.0
Lead	—	0.05
Mercury	—	.002
Nitrate	—	010
Selenium	—	0.01
Silver	—	0.05
Organic Contaminants		
2,4-D	—	0.1
Endrin	—	0.0002
Lindane	—	0.004
Methoxychlor	—	0.1
2,4,5-TP Silvex	—	0.01
Benzene	0	0.005
Carbon tetrachloride	0	0.005
p-Dichlorobenzene	0.075	0.075
1,2 - Dichloroethane	0	0.005
1,1 - Dichloroethylene	0.007	0.007
1,1,1 - Trichloroethane	0.20	0.20
Trichloroethylene	0	0.005
Vinyl chloride	0	0.002
Total trihalomethanes	—	0.10
(Chloroform, bromoform, bromodichloromethane, dibromochloromethane)		
Radionuclides		
Gross alpha particle activity	—	15 pCi/L
Gross beta particle activity	—	4 mrem/yr
Radium 226 and 228 (total)	—	5 pCi/L

CLGs -
SMCLs -   Maximum Contaminant Levels
HPC - heterotrophic plate count
pCi/L - picocuries per liter
(Courtesy, U.S. EPA, 1980)

This entire matter is constantly under study, primarily because of the implementation of new amendments to the Safe Water Drinking Act (SWDA, 1974). As a result, it would be prudent that those concerned with the production of a potable water keep informed of the extensive and pending requirements. Note that the parameters at issue are intended to supply a water that is pleasing, suitable for use, and safe.

Recommended parameters. The substances that are shown under "Recommended" are such that a concentration higher than those listed, while not constituting a health hazard, might affect the suitability of this water for the following reasons:

Chloride (Cl) - Salty taste, corrosion.
Copper (Cu) - Astringent taste, discoloration of fixtures, corrosion.
Iron (Fe) - Taste, discoloration, staining of laundry and fixtures.
Manganese (Mn) - Taste, discoloration, staining of laundry and fixtures.
Sodium (Na) - Taste, potential health consequences.
Sulfate (SO_4) - Gastrointestinal irritation when combined with Mg
                    (forms epsom salts) or sodium.
Zinc (Zn) - Astringent taste.
Total dissolved solids (TDS) - Taste, gastrointestinal irritation.
Color - Aesthetics, indicates other substances (e.g., Fe, Mn, Cu, organics).
pH - Corrosion and taste.
Foaming agents (MBAS) - Aesthetic problems.

Not included in these standards are limits for calcium, magnesium, or bicarbonates, although these exist in almost every water supply. There is no evidence that any of these ions will affect the color, odor, taste, or appearance of water; but calcium and magnesium do cause hardness that causes excessive scaling of pipes, water heaters, and other equipment.

It is noticed that the World Health Organization does include parameters for calcium and magnesium because of the hardness factor. Further, a certain concentration of magnesium sulfate ($MgSO_4$, epsom salts) can result in gastrointestinal irritation.

Maximum contaminant levels (MCL). In 1986 a set of amendments to the Safe Drinking Water Act (1974) was legislated. These have been adopted and will most likely be implemented in the future. The major regulations pertain to lead, copper, disinfection by-products (DBP), and also to other organic compounds.

If the parameters as listed above are exceeded, the following toxic effects are forecast:

Table 1-6
Constituents in a Typical Water Supply

Class 1	*Primary Constituents — Generally over 5 mg/L*

Bicarbonate	Magnesium	Sodium
Calcium	Organic matter	Sulfate
Chloride	Silica	Total dissolved solids

Class 2	*Secondary Constituents — Generally over 0.1 mg/L*

Ammonia	Iron	Potassium
Borate	Nitrate	Strontium
Fluoride		

Class 3	*Tertiary Constituents — Generally over 0.01 mg/L*

Aluminum	Copper	Phosphate
Arsenic	Lead	Zinc
Barium	Lithium	
Bromide	Manganese	

Class 4	*Trace Constituents — Generally less than 0.01 mg/L*

Antimony	Cobalt	Tin
Cadmium	Mercury	Titanium
Chromium	Nickel	

Class 5	*Transient Constituents*

Acidity/alkalinity
Biological cycles:
  Organic — $C/CH_4/CO/CO_2/(CH_2O)_n$
  Oxygen — O_2/CO_2
  Nitrogen — $N/NH_3/NO_2/NO_3/N_2$/Amino acids
  Sulfur — $S/H_2S/SO_3/SO_4$
Oxidants: $O_2/S/CL_2/CrO_4$
Reducing agents: Organics/$Fe^{2+}/Mn^{2+}/H_2S/SO_2/SO_3$
Radionuclides

Nonsoluble constituents

Floating solids	Settleable solids	
Suspended solids	Algae	
Bacteria	Fungi	Viruses

(Courtesy, U.S. Geological Survey)

Arsenic (As) - Damage to nervous system, toxicity effects.
Fluoride (F) - Skeletal damage, mottling and disfiguration of teeth.
Lead (Pb) - Damage to nervous system, kidney effects, toxic to infants.
Mercury (Hg) - Damage to central nervous system, kidney effects.
Nitrate (NO$_3$) - Oxygen deprivation in infants, methemoglobinemia.
Selenium (Se) - Damage to nervous system.
Turbidity - Interferes with disinfection, possible gastrointestinal irritation.
Coliform bacteria - Indicative of fecal contamination; and when present in excess of parameter indicates that other disease-producing organisms may be present.
Total trihalomethanes (TTHM) - All such organic compounds are a potential cancer risk (carcinogenic).

There is little likelihood that river, lake, or well sources are capable of delivering an acceptable product water without some treatment. Municipalities and private purveyors are the sources for many manufacturing plants whose requirements range from supplying a low-pressure boiler, to the high-purity water required for the semiconductor, pharmaceutical, and nuclear energy industries. Considering the mandated regulations under which municipalities must operate, it might be assumed that municipal supplies could satisfy some of these demands.

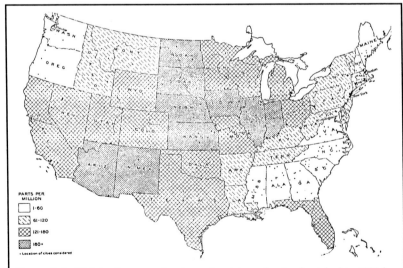

Figure 1-5. Weighted average hardness, by states, of water furnished in 1932 by public supply systems in over 600 cities in the United States.
(Courtesy, U.S. Geological Survey)

Such is not the case. The municipalities are required to remove only those impurities that make the drinking water unpalatable and unsafe. They cannot, as a matter of practicality and/or cost, deliver an all-purpose water that is suitable for everything from general home consumption to all the many other uses industries have for water. As a result, although great strides have been made in the quality of municipal supplies, these water sources do not meet the parameters of most industries.

As a case in point: People still ingest more water than they do soft drink beverages; and yet it is safe to say that practically 100% of these bottling plants include water treatment equipment, even though most of them, not only in the United States but throughout the world, use municipal water.

Not all beverage plants require the same type of water treatment. Some plants require alkalinity reduction, some plants require removal of organic substances, and all plants require removal of all chlorine or chlorinated compounds.

Thus it can be noted that although the municipal supply and soft drinks are both very refreshing beverages, the municipal supply does not meet the parameters required for a first-class soft drink. (Requirements for beverage water and other product waters are further discussed under "Product Waters," Chapter 15.)❑

CHAPTER 2

STRUCTURE OF MATTER

Almost everyone is aware of the fact that H_2O is the chemical formula for water. To the chemist, however, this representation signifies a somewhat more complex phenomenon, bringing to mind *elements, atoms, ions, molecules,* and *compounds.* The definitions of these words are straightforward, but since several of them are used synonymously, they are clarified in the sections that follow.

Elements

An element is a substance that cannot be decomposed without losing its unique chemical or physical properties. An element can be broken down by nuclear disintegration into two or more other substances, but the original element will no longer exist. There are over one hundred elements known today, and of these less than twenty are normally encountered in water and water treatment. Should the water contain other elements, they would be of such rarity and/or quantity that their effects would be negligible, or would require special handling.

The elements, and ions containing several different elements, shown in Table 2-1 make up approximately 95% of the earth's crust.

Atoms and Ions

Matter is made up of many atoms. Each atom is the smallest unit of matter that

Table 2-1
Elements in Earth's Crust

Metals	Nonmetals	Ions
Calcium (Ca)	Oxygen (O)	Bicarbonate (HCO_3^-)
Magnesium (Mg)	Hydrogen (H)	Carbonate (CO_3^{2-})
Sodium (Na)	Carbon (C)	Hydroxide (OH^-)
Potassium (K)	Chlorine (Cl)	Sulfate (SO_4^{2-})
Iron (Fe)	Sulfur (S)	Nitrate (NO_3^-)
Aluminum (Al)		

retains the characteristics of an element. An atom is made up of a *nucleus* containing one or more protons, which carry a positive electrical charge; and (except for hydrogen) one or more *neutrons*, which carry no charge. The nucleus is orbited by sufficient negatively charged electrons to neutralize its positive charge.

Figure 2-1 shows several simple atoms: hydrogen, helium, and lithium. Other atoms contain greater numbers of protons, neutrons, and electrons; and are more complex structures. At times this arrangement of the neutral atom is upset. Natural or induced influences may knock another atom's original electron(s) out of natural orbit(s). When this occurs, the affected atom is no longer neutral. This particular atom is now called an *ion*, which is defined as an electrically charged atom.

For example, the hydrogen atom depicted in Figure 2-1 contains one positive charge (a proton) and one negative charge (an electron). If the negative charge is lost, this atom becomes positively charged and is now called a *cation*. On the other hand, an atom such as chlorine will pick up an electron to make the chlorine atom negatively charged. In this case the altered atom is called an *anion*.

The concept of these charges and their behavior can be visualized by comparing them to the familiar charges of a magnet. A magnet is a piece of iron or steel that will attract or repel other pieces of iron or steel. Magnets can be naturally occurring magnetic substances (magnetite); or magnetic properties can be induced with an electric current on certain nonmagnetic substances. Whichever the case, the forces of magnetism exist because of the special alignment of the positive and negative charges that exist in the magnetic material, as shown in Figure 2-2.

When two magnets are placed end to end they will repel or attract each other, as shown in Figure 2-3.

It becomes obvious from Figure 2-3 that similar charges repel each other and opposite charges will attract each other. This behavior in cations and anions is the subject of this study.

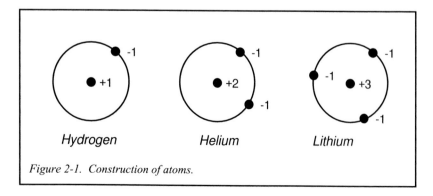

Figure 2-1. Construction of atoms.

Returning to the magnet, remember that it contains millions of atoms. When the forces of the individual atoms are combined, a magnet of considerable strength will exist. The ionic character and strength of some of the atoms listed in Table 2-1 are shown in Table 2-2:

Valence
Valence is the term used to express ionic strength, and it can be further defined as the *combining power of the various ions.* The valences of various ions are indicated in Table 2-2. For example, the valence for calcium is plus two (2+), and that of oxygen is minus two (2-). Ions are the building blocks of chemistry, and their behavior dictates both the impurities (compounds) in the water and the requirements for the removal of those compounds.

Since similar charges repel each other, there can be no compounds such as $Ca^{2+}Mg^{2+}$ or $OH^-HCO_3^-$. However, the cations and the anions will or can attract each other, to result in compounds such as Na^+Cl^- and H^+OH^-. Plus and minus charges are of equal intensity. A 1+ ion will combine with a 1- ion, a 2+ ion will combine with two 1- ions, and a 2- ion will combine with two 1+ ions. Following are examples of the combining of ions of different charges and valences:

1+ and 1-
$Na^+ + Cl^- = NaCl$ (sodium chloride)
$H^+ + OH^- = HOH$ (water)
$2H^+ + O^{-2} = H_2O$ (water)

2+ and 1-
$Ca^{2+} + 2Cl^- = CaCl_2$ (calcium chloride)
$Mg^{2+} + 2HCO_3^- = Mg(HCO_3)_2$ (magnesium bicarbonate)

1+ and 2-
$2Na^+ + SO_4^{2-} = Na_2SO_4$ (sodium sulfate)
$2H^+ + O^{2-} = H_2O$ (water)

Figure 2-2. In the bar on the left, electrical charges are at random; thus no magnet exists. When the charges are aligned (bar on the right), a magnet exists.

2+ and 2-
$$Ca^{2+} + SO_4^{2-} = CaSO_4 \text{ (calcium sulfate)}$$
$$Mg^{2+} + SO_4^{2-} = MgSO_4 \text{ (magnesium sulfate)}$$

3+ and 2-
$$2Al^{3+} + 3SO_4^{2-} = Al_2(SO_4)_3 \text{ (aluminum sulfate)}$$

3+ and 1-
$$Al^{3+} + 3OH^- = Al(OH)_3 \text{ (aluminum hydroxide)}$$

Molecules
A molecule is a combination of atoms as shown in the compounds listed above, and is the smallest particle of matter that will retain the characteristics of a compound. The discussion of the combining of ions may sound somewhat academic, but it is the basis for the interactions of all compounds, including those encountered in the treatment of water.

There are, of course, several other compounds composed of those ions shown in Table 2-1 and 2-2 that are found in water. This is not to say that these are not of importance, but for the sake of simplicity, they will be omitted here and will be discussed if and when required.

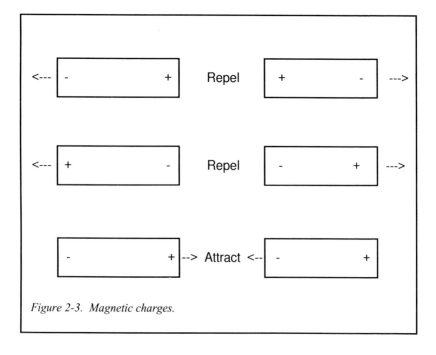

Figure 2-3. Magnetic charges.

Hypothetical Combinations[*]

The complexity of complete testing of water is such that analysis must be performed in a laboratory. The lab does not usually determine the quantities of the compounds listed in Table 2-3, but it does determine the absolute weights of the ions that make up the compounds.

In other words, the lab reports the actual weights of the cations (such as Ca^{2+}, Mg^{2+}, and Na^+) and the weights of the anions (such as HCO_3^-, SO_4^{-2}, and Cl^-) that are in the water. It then remains for those who use the lab report to determine the compounds that might exist in the water.

To this end, a method has been devised for determining probable hypothetical combinations of ions in water. The procedure below is self-explanatory, but its use will be detailed in Chapter 4.

The order of precedence for the combining of ions in water is as follows:

Cation	Anions
1. Ca	1. HCO_3
2. Mg	2. SO_4
3. Na	3. Cl

The first two ions that combine would be Ca and HCO_3. If any calcium remains, the second combination would be Ca and SO_4. In order of probability, a complete listing of the possible combinations follows:

[*]Also see Chapter 4, "Interpretation of Chemical Analyses."

Table 2-2
Ionic Character and Strength

Cations	Anions
Calcium (Ca^{2+})	Oxide (O^{2-})
Magnesium (Mg^{2+})	Chloride (Cl^-)
Sodium (Na^+)	Bicarbonate (HCO_3^-)
Potassium (K^+)	Carbonate (CO_3^{2-})
Iron (Fe^{2+}, Fe^{3+})	Hydroxide (OH^-)
Manganese (Mn^{2+}, Mn^{3+})	Nitrate (NO_3^-)
Hydrogen (H^+)	Sulfate (SO_4^{2-})
Aluminum (Al^{3-})	

Note: The charges of ions are either positive or negative, and the number of charges varies from one to five.

1. Any Ca will first combine with HCO_3.
2. If Ca remains, it will combine with SO_4.
3. If Ca still remains, it will combine with Cl.
4. *If OH exists, it will combine first with Mg.
5. *If OH remains, it will combine with any Ca or any Na that remains.
6. If HCO_3 still remains, it will combine with Mg.
7. If any HCO_3 remains, it will combine with Na.
8. If any Mg is left over from step 4, it will combine with SO_4.
9. If any Mg remains, it will combine with Cl.
10. Any remaining SO_4 from step 8 will combine with Na.
11. Any chloride remaining from step 9 will combine with Na.
12. If NO_3 (nitrates) exist, they will combine with any Ca, Mg, or Na in that order. (Nitrates can exist if there is a polluted source.)❑

*Chances are that the raw water supply to the plant will seldom, if ever, contain hydroxides (OH). These are listed here for use when lime or sodium hydroxide is used in the treatment of the plant supply.

Table 2-3
Prevalent Compounds in Water

$Ca(HCO_3)_2$	Calcium bicarbonate
$Ca(OH)_2$	Calcium hydroxide
$CaCO_3$	Calcium carbonate
$CaSO_4$	Calcium sulfate
$CaCl_2$	Calcium chloride
$Mg(HCO_3)_2$	Magnesium bicarbonate
$Mg(OH)_2$	Magnesium hydroxide
$MgCO_3$	Magnesium carbonate
$MgSO_4$	Magnesium sulfate
$MgCl_2$	Magnesium chloride
$NaHCO_3$	Sodium bicarbonate
$NaOH$	Sodium hydroxide
Na_2CO_3	Sodium carbonate
Na_2SO_4	Sodium sulfate
$NaCl$	Sodium chloride

Complete table in appendix

CHAPTER 3

MEASUREMENTS AND CALCULATIONS

Mass can be defined as "how heavy things are." When used alone, this term indicates the actual (absolute) mass, which is measured in units such as ounces, pounds, or grams. However, use of the unit of mass alone is not common in water chemistry. Rather, the following units are used:

1. Mass of substance per mass of water, and/or
2. Mass of substance per volume of water.

These units are called *concentrations*, and they denote the strength of solutions. The word *mass* is also used in connection with other characteristics such as *atomic mass* and/or *molecular mass*. When used in this sense, however, mass is not measured by ounces, pounds, grams, or any other commonly used standard. The most effective way to explain these terms is to discuss each in depth.

Concentrations

Why is it necessary to resort to such parameters as parts per million (ppm) or milligrams per liter (mg/L)?

The following example demonstrates the necessity: The permissible level of total dissolved solids in potable water is 0.00417 pounds per gallon (lb/gal) (4.17×10^{-3} lb/gal). This is equivalent to 0.0667 ounces per gallon (oz/gal) (6.67×10^{-2} oz/gal).

These numbers indicate that the quantity of dissolved solids is relatively a very small amount compared to the substance in which they are dissolved. Further, these numbers are not only cumbersome with which to work, but their use would be prone to error. Compare 0.0667 oz/gal with 500 ppm, which is an equivalent concentration, and the advantage can be seen.

Parts per million (ppm). One part per million equals one pound of any substance in one million pounds of water. One part per million equals one ounce of any substance in one million ounces of water. In other words, one part per

Table 3-1
Conversion of U.S. to Other Units

Units	Parts CaCO₃ per Million (ppm)	Grains CaCO₃ per U.S. Gallon (gr/gal)	English Degrees or Clark	French Degrees	German Degrees	Milliequivalents per Liter (eq/million)
1 part per million	1.	0.0583	0.07	0.1	0.0560	0.020
1 grain per U.S. gallon	17.1	1.	1.2	1.71	0.958	0.343
English or Clark degree	14.3	0.833	1.	1.43	0.800	0.286
1 French degree	10.	0.583	0.7	1.	0.560	0.20
1 German degree	17.9	1.04	1.24	1.79	1.	0.357
1 milliequivalent/L*	50.0	2.92	3.50	1.79	2.80	1.

*Note: Definition of the milliequivalents/liter unit and a discussion is found at the end of Chapter 3.

Table 3-2
Conversion Units and Equivalents

Water Analysis Units	Parts per Million (ppm)	Milligrams per Liter (mg/L)	Grams per Liter (g/L)	Grains U.S. Gallon (gr/U.S. gal)	Grains British Imp. Gal	Kilograins per Cubic Foot (kgr/ft³)
1 part per million	1.	1.	0.001	0.583	0.07	0.0004
1 milligram per liter	1.	1.	0.001	0.583	0.07	0.0004
1 gram per liter	1,000.	1,000.	1.	68.3	70.0	0.436
1 grain per U.S. gallon	17.1	17.1	0.017	1.	1.2	0.0075
1 grain per British imp. gal	14.3	14.3	0.014	0.833	1.	0.0062
1 kilograin per cubic foot	2,294.	2.294	229.4	134.	161.	1.

1 mg/L = 1 ppm 10⁻³ g/L   1 microgram/liter (μg/L) = 1 part per billion (ppb) 10⁻⁶ g/L
1 nanogram/liter (ng/L) = 1 part per trillion (ppt) 10⁻⁹ g/L

million is equal to one part of any substance in one million similar parts of water. By way of comparison in gallons, 1 ppm equals 1 lb of substance in 120,048 gal of water.

Milligrams per liter (mg/L). 453.6 grams (g) = 1 lb
1 g = 1,000 milligrams (mg)
1 gal = 3.785 liters (L)
1 gal = 3,785 milliliters (mL)

By definition, the only difference between parts per million and milligrams per liter is one of water temperature. For mg/L, the temperature of the water is standardized at approximately 4 °C (40 °F). The variance of volumes of 1 liter of water at various temperatures is negligible, however, and this stipulation is generally disregarded. For our purposes, it is assumed that ppm and mg/L are of the same value.

Grains per gallon (gr/gal). One pound equals 7,000 grains. As a rule, 1 gr/gal (which is equal to 17.1 ppm) is used to identify chemical dosage and hardness of water, and to report the capacities of ion-exchange resins.

Although there is an interchangeability of these various units, it has been the author's experience that ppm is still the most widely used, and this will likely be the case until complete conversion to the metric system has been established.

Tables 3-1 and 3-2 contain frequently used conversions, and it is suggested that one become familiar with them.

Atomic Mass

The atom of each element has its own specific mass that is different from that of the atom of any other element. That mass is called that element's atomic mass. When the atomic theory was first propounded, there was no method available to determine the actual (absolute) mass of an atom; however, it was possible to observe the relative weights of atoms.

It was known that hydrogen is the lightest atom, so it was given the atomic mass of one (1.0). Through experimentation it was learned that oxygen is approximately 16 times as heavy as hydrogen, so oxygen was given the atomic mass of 16. In a like manner, carbon was given an atomic mass of 12. These masses are therefore called relative masses, and have no relationship to actual weights such as ounces, pounds, or grams. Some of the common atomic masses are given in Table 3-3, and a complete list is given in Appendix .

Molecular Weight

The molecular weight is equal to the sum of the atomic masses that make up the molecule. For example: the atomic mass of sodium is 23.0, the atomic mass of chlorine is 35.5; thus the molecular mass of sodium chloride is 58.5. Table 3-4 compares the molecular weights of some common substances.

Useful Calculations
Some useful equivalents are given in Table 3-5, followed by examples of calculations of other equivalents.

Following is an example of how these units can be used to develop other equivalents: 1 lb/1,000,000 lb = 1 ppm, and since 1 gal of water weighs 8.33 lb, then 1,000,000 lb of water has a volume of 1,000,000/8.33 = 120,048 gal. Therefore 1 lb/120,048 gal = 1 ppm.

Now, if both sides of the equations are multiplied by 120,048:
$$\frac{1 \text{ lb} \times 120,048}{120,048 \text{ gal}} = 1 \text{ ppm} \times 120,048$$

1 lb/1,000 gal = 120,048 ppm or 1 lb/1,000 gal = 120 ppm ±

Table 3-3
Atomic Masses of Common Elements and Ions

Aluminum (Al)	26.97
Calcium (Ca)	40.08
Carbon (C)	12.010
Chlorine (Cl)	35.457
Copper (Cu)	63.57
Hydrogen (H)	1.008
Iron (Fe)	55.85
Magnesium (Mg)	24.32
Nitrogen (N)	14.008
Oxygen (O)	16.00
Potassium (K)	39.096
Silver (Ag)	107.88
Sodium (Na)	22.99
Zinc (Zn)	65.38
Bicarbonate (HCO_3)	61.018
Carbonate (CO_3)	60.01
Hydroxide (OH)	17.00
Nitrate (NO_3)	62.008
Sulfate (SO_4)	96.06

(Complete table in appendix.)

Another example:
   Plant capacity = 10,000 gallons per hour (gph)
   10,000 × 8 h = 80,000 gallons per day (gpd)

Assume that the amount of chemical determined from laboratory tests is 2 gr/gal. (This is called the *dosage*.)
   80,000 gpd × 2 gr/gal = 160,000 gr/day

Then 160,000 gr/day at 7,000 gr/lb = 22.8 lb of chemicals/day

Percent (%)
The word *percent* is derived from the Latin *per* (by) and *centum* (hundred). In other words, it indicates the parts per hundred: 25% of 100 = 25; 50% of 100 = 50; and 5% of 100 = 5. For calculating purposes, 5% is written as 0.05, 25% as 0.25, and 50% as 0.50: 0.05 × 100 = 5; 0.25 × 100 = 25; and 0.50 × 100 = 50. Also 5% of 50 = 0.05 × 50 = 2.5; 25% of 50 = 0.25 × 50 = 12.5; and 50% of 50 = 0.50 × 50 = 25.

An example of how an operator might use percentages is in solving the following problems:

(A) What is the percent of the solution when 1 lb of chlorine is mixed in 1 gal of water?

Table 3-4
Molecular Weights of Common Substances

Calcium bicarbonate	$(Ca(HCO_3)_2)$	162.1
Calcium hydroxide	$(Ca(OH)_2)$	74.1
Calcium carbonate	$(CaCO_3)$	100.8
Calcium sulfate	$(CaSO_4)$	136.1
Calcium chloride	$(CaCl_2)$	111.0
Magnesium bicarbonate	$(Mg(HCO_3)_2)$	146.3
Magnesium hydroxide	$(Mg(OH)_2)$	58.33
Magnesium carbonate	$(MgCO_3)$	84.3
Magnesium sulfate	$(MgSO_4)$	120.4
Magnesium chloride	$(MgCl_2)$	95.2
Sodium bicarbonate	$(NaHCO_3)$	84.0
Sodium hydroxide	$(NaOH)$	40.0
Sodium carbonate	(Na_2CO_3)	106.0
Sodium chloride	$(NaCl)$	58.5

$$\frac{1 \text{ lb of chlorine } (100\%)}{1 \text{ lb of chlorine} + 8.33 \text{lb (lb of water in 1 gal)}} = \% \text{ by mass}$$

$$\frac{1}{9.33} = 0.107 \text{ or } 10.7\% \text{ by mass}$$

(Note: To convert 0.107 to 10.7%, move the decimal two places to the right.)

(B) How much 70% available chlorine powder is required to give 5 gal of 10% solution?

Let y = lb of 70% chlorine powder

$$\frac{0.7y}{0.7y + (8.33 \text{ lb/gal} \times 5)} = 0.10$$

$$0.7y = 0.10 \times (0.7y + 41.65 \text{ lb})$$
$$0.7y = .07y + 4.165 \text{ lb}$$
$$0.63y = 4.165$$
$$y = \frac{4.165}{0.63} = 6.6 \text{ lb of } 70\% \text{ Cl}_2$$

Table 3-5
Useful Equivalents

1 gal of water weighs approximately 8.33 lb
1 gal = 231 cubic inches (in.^{3})
1 gal = 3.785 L
1 gal = 3,785 mL
1 cubic foot (ft^3) of water weighs approximately 62.4 lb
1 ft^3 = 7.48 gal
1 lb = 7,000 gr
1 lb = 453.6 g
1 g = 15.43 gr
1 g/L = 58.41 gr/gal
1 gr/L = 1,000 ppm
0.0171 g/L = 1 gr/gal
0.0038 g/gal = 1 ppm
1 gr/gal = 17.1 ppm
1 gr/gal = 142.9 lb/1,000,000 gal
1 mg/L = 1 ppm
1 ppm = 0.058 gr/gal
8.33 lb/1,000,000 gal = 1 ppm
10,000 ppm = 1%

(C) How many gallons of 12% chlorine solution are required to make up 40 gal of 5% solution?

(1) Forty gal of 5% solution will contain 40×8.33 lb/gal $\times$ $0.05 = 16.66$ lb of chlorine (100%)

(2) One gal of 12% solution contains 1 gal $\times 8.33$ lb/gal $\times 0.12 = 0.996$ lb of chlorine (100%) per gallon.

(3) To obtain 16.66 lb of chlorine, it will be necessary to use $16.66 \div 0.996 = 16.72$ gal of 12% solution.

(4) Therefore, 16.72 gal of 12% chlorine solution must be diluted to 40 gal.

(D) If it is desired to make up 10 gal of a 50-ppm solution using a 5% solution of chlorine.

$$50 \text{ ppm} \div 17.1 = 2.92 \text{ gr/gal}$$

Therefore, 10 gal would contain 10 gal $\times 2.92$ gr/gal $= 29.2$ gr.

1 gal of 5% solution contains:

$$8.33 \text{ lb/gal} \times 0.05 = 0.4165 \text{ lb of } 100\% \text{ } Cl_2 \text{ and}$$
$$0.4165 \text{ lb} \times 7,000 \text{ gr/lb} = 2,915.5 \text{ gr}$$

Since 1 gal contains 3,785 cubic centimeters (cm^3), each cm^3 of above contains:

$$\frac{2,915.5}{3,785} = 0.7703 \text{ gr/cm}^3$$

Therefore,

$$\frac{29.2 \text{ gr}}{0.7703 \text{ gr/cm}^3} = 37.9 \text{ cm}^3 \text{ of 5\% solution}$$

will be required in 10 gal of water to give a solution of 50 ppm.

Efficiency of Removal. If 5 ppm of turbidity is removed from a sample of raw water that contains 70 ppm of turbidity, what is the efficiency of removal?

$$\frac{\text{ppm of turbidity removed}}{\text{ppm of turbidity in raw water}} = \% \text{ efficiency}$$

$$5 \div 70 = 0.07 = 7\%$$

(Note: To convert 0.07 to 7%, move the decimal two places to the right.)

This can also be shown as follows:

$$\frac{\text{ppm turbidity in raw water - ppm turbidity in treated water}}{\text{ppm turbidity in raw water}} = \text{efficiency removal}$$

$$\frac{(70 - 65)}{70} = \frac{5}{70} = 0.07 = 7\%$$

Mixtures

There will be times where it is necessary to produce a mixture of a certain strength from two different products (for example, when it is desired to get a mixture of a product of 50% strength from one product that has a concentration of 60% and another product that has a concentration of 30%). Rather than to resort to the algebraic methods, the rectangle method can be used. Thus, by mixing 10 parts of the 30% concentration and 20 parts of the 60% concentration, a mixture of 30 parts of 50% concentration will result.

Temperature Conversions

To convert centigrade degrees to Fahrenheit degrees, multiply the number of centigrade degrees by 9/5 and add 32.

Example: To convert 100 °C to °F,

$$°C = 100 \times 9/5 + 32 = 180 + 32 = 212 \text{ °F}$$
$$100 \text{ °C} = 212 \text{ °F}.$$

To convert Fahrenheit degrees to centigrade degrees, subtract 32 from the number of Fahrenheit degrees and multiply by 5/9.

Example: To convert 212 °F to °C,

$$(212 - 32) \times 5/9 \text{ °F} = 180 \times 5/9 = 100 \text{ °C}$$
$$212 \text{ °F} = 100 \text{ °C} \quad \square$$

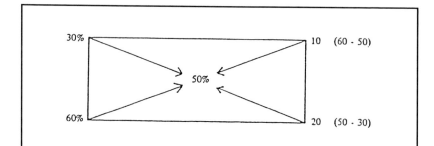

Figure 3-1. The rectangle method of combining two concentrations to get a different concentration.

CHAPTER 4

INTERPRETATION OF CHEMICAL ANALYSIS

A typical analysis (Table 4-1) of an untreated plant water supply will give information as to the principal elements of that water.

Although the quantity of each element is reported in the same unit (ppm), each carries different *relative weights* (see Chapter 2, "Structure of Matter"). Note, for example, the result when the positive and negative ions from Table 4-1 are combined:

Ca^{2+}	32.0	HCO_3^-	121.9
Mg^{2+}	7.3	SO_4^{2-}	19.2
Na^+	9.2	Cl^-	7.1
	48.5		148.2

The imbalance indicates that the above exercise was one of adding apples and oranges; hence it becomes apparent that the individual values must be correlated.

Equivalent Weight

In order to be able to correlate these values, they must be referred to and converted to a standard or common denominator. In water treatment the standard unit is calcium carbonate ($CaCO_3$) equivalence and the reason is really one of fortunate convenience. *The molecular weight of calcium carbonate is 100.*

Since each of the elements has a different atomic weight, it can be converted to its calcium carbonate equivalent, and it will be noted that each element has its own conversion factor (Table 4-2). It is determined as follows:

(molecular weight of $CaCO_3$) ÷ atomic weight of element = equivalent weight

To determine the conversion constant of calcium, for example:

$$100 \div 40 = 2.5$$

This is so because the valence of both $CaCO_3$ and Ca is 2. However, if the valence of an element is 1 (e.g., Na), then the calculation is as follows:

$$(100 \div 23) \times 1/2 = 2.17$$

The same can be said of a compound. Consider $Fe(OH)_3$ in which the valence of Fe = 3 and whose molecular weight = 107; then:

$$CaCO_3 \text{ equivalent} = 100 \div 107 \times 3/2 = 1.41$$

In a like manner, a conversion constant can be developed for each of the elements and compounds with which this chemistry is involved.

In order to make this determination, it is first necessary to convert each of the above to its $CaCO_3$ equivalent. Refer to the conversion Table 4-2 and make the following calculations:*

Constituent*	as $CaCO_3$	
Ca (32.0 ppm) × 2.5	= 80.0 ppm	total cations
Mg (7.3 ppm) × 4.1	= 30.0 ppm	130.0
Na (9.2 ppm) × 2.17	= 20.0 ppm	
HCO_3 (121.9 ppm) × 0.82	= 100.0 ppm	total anions
SO_4 (19.2 ppm) × 1.04	= 20.0 ppm	130.0
Cl (7.1 ppm) × 1.40	= 10.0 ppm	

Hypothetical Combinations

Since all the values have been converted to a common denominator ($CaCO_3$), it is now possible to determine the proportions and combinations in which they exist.

The rule to be followed is that *bicarbonates* will first combine with calcium to give $Ca(HCO_3)_2$. If more bicarbonates exist than calcium, the remainder of the bicarbonates will then combine with magnesium to give $Mg(HCO_3)_2$. Should any bicarbonates still exist, they will combine with sodium to give $NaHCO_3$ (see Chapter 2). The same is true of the remaining anions and cations.

The order of preference for combining, you may recall, is as follows:

Cations	*Anions*
1. Ca	1. HCO_3
2. Mg	2. SO_4
3. Na	3. Cl

Following the order of preference inChapter 2:

---

*This analysis was designed so that all equivalent values would be whole numbers. This probably never occurs in actual practice, but is done here so that there will be no distraction from the important topic.

1. Eighty ppm of Ca will combine with 80 ppm of HCO_3 to give <u>80 ppm $Ca(HCO_3)_2$</u>. This leaves 20 ppm of HCO_3.

2. No Ca remains to combine with SO_4.

3. No Ca remains to combine with Cl.

4. No OH exists*.

5. No OH exists*.

6. Twenty ppm of Mg will combine with 20 ppm of HCO_3 to give <u>20 ppm of $Mg(HCO_3)$</u>.

This leaves no HCO_3 and 10 ppm of Mg.

7. No HCO_3 remains.

8. Ten ppm of Mg will combine with 10 ppm of SO_4 to give <u>10 ppm of $MgSO_4$</u>.

This leaves no Mg and 10 ppm of SO_4.

Table 4-1 Typical Water Analysis	
Physical Properties	
Turbidity	0.0
Color	5.0
Chemical Properties	*ppm*
Calcium (Ca)	32.0
Magnesium (Mg)	7.3
Sodium (Na)	9.2
Bicarbonates (HCO_3)*	121.9
Sulfates (SO_4)	19.2
Chlorines (Cl)	7.1
Carbon dioxide (CO_2)	
Oxygen consumed	
pH	7.1
Phosphates (PO_4)	
Nitrites (NO_2)	
Nitrates (NO_3)	
Iron (Fe)**	

*Sometimes reported as $(CaCO_3)$. If so, no conversion required.
**All Fe and Mn should be removed during treatment, so these values are not considered now.

9. The remaining 10 ppm of SO_4 will combine with 10 ppm of Na to give <u>10 ppm Na_2SO_4</u>.

This leaves 10 ppm of Na and 10 ppm of Cl.

10. No SO_4 remains.

11. The 10 ppm of Cl will combine with 10 ppm of Na to give <u>10 ppm NaCl</u>.

12. No NO_3.

*In untreated water, be it from wells or surface, OH is not likely to exist; however, OH can be found in those waters after treatment with lime ($Ca(OH)_2$) or caustic (NaOH).

Table 4-2
Calcium Carbonate Equivalents Conversion Factor

To determine the calcium carbonate equivalent of the following:	Multiply by:	To determine the calcium carbonate equivalent of the following:	Multiply by:
Ca^{+2}	2.50	NaCl	0.856
CaO	1.79	Na_2SO_4	0.705
$Ca(OH)_2$	1.35	$NaNO_3$	0.588
$Ca(HCO_3)_2$	0.617	K^+	1.28
$CaCO_3$	1.00	K_2O	1.06
$CaCl_2$	0.902	KOH	0.89
$CaSO_4$	0.735	$KHCO_3$	0.499
$Ca(NO_3)_2$	0.610	K_2CO_3	0.724
KCL	0.67		
Mg^{2+}	4.12	K_2SO_4	0.574
MgO	2.48	KNO_3	0.495
$Mg(OH)_2$	1.72		
$Mg(HCO_3)_2$	0.684	OH^-	2.94
$MgCO_3$	1.19	CO_3^{2-}	1.667
$MgCl_2$	1.05	HCO_3	0.820
$MgSO_4$	0.831	Cl^-	1.41
$Mg(NO_3)_2$	0.674	SO_4	1.04
SO_3	1.25		
Na^+	2.17	SO_2	1.56
Na_2O	1.61	NO_3	0.805
$NaHCO_3$	0.596	H_2SO_4	1.02
Na_2CO_3	0.944	HCL	1.37

The cations (Ca, Mg, and Na) cannot exist alone. They exist in combination with the anions (HCO_3, SO_4, Cl), and the probable manner in which they exist is not usually shown in the analysis.

The resulting analysis has been developed as follows:

$Ca(HCO_3)$	= 80.0 ppm (as $CaCO_3$)
$CaSO_4$	= 0.0 ppm " "
$CaCl_2$	= 0.0 ppm " "
$Mg(HCO_3)_2$	= 20.0 ppm " "
$MgSO_4$	= 10.0 ppm " "
$MgCl_2$	= 0.0 ppm " "
$NaHCO_3$	= 0.0 ppm " "
Na_2SO_4	= 10.0 ppm " "
NaCl	= 10.0 ppm " "
Total dissolved solids	= 130.0 ppm (as $CaCO_3$)

The same results can be obtained graphically if drawn with care (Figure 4-1):

Note that both the cations and anions equal 130 ppm. Again be reminded that such exact balance is unlikely in actual practice.

After a familiarity is developed as to which compounds exist in water, it is time to determine their properties, and what effect they will have on the quality of the water. These characteristics are discussed in Chapter 5, "Basic Parameters."

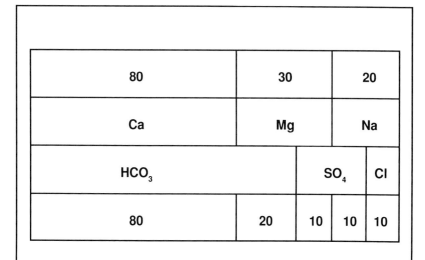

Figure 4-1. Graph of analysis.

Milliequivalents per Liter (meq/L)

Using the same analysis previously under discussion, a second method for reporting the concentration of ions is described. This method has been seen more frequently of late and may eventually become the universal standard.

Ion	ppm
Ca	32.0 ppm
Mg	7.3 ppm
Na	9.2 ppm
HCO_3	121.9 ppm
SO_4	19.2 ppm
Cl	7.1 ppm

In this method, the numerical value of 1 meq/L is equal to the atomic weight of the ion divided by its valence. For example:

$$1 \text{ meq/L Ca} = 40 \div 2 = 20 \text{ mg/L (20 ppm)}$$
$$2 \text{ meq/L Ca} = 40 \text{ mg/L (40 ppm)}$$

Table 4-3
A Partial List of Conversion Factors to meq/L

Substance	meq/L
Ca	20.0
$CaCO_3$	50.0
$CaHCO_3$	80.0
$CaSO_4$	68.0
$CaCl_2$	55.5
$Ca(OH)_2$	37.0
Mg	12.1
$MgCO_3$	42.0
$MgHCO_3$	73.0
$MgSO_4$	60.0
$MgCl_2$	47.5
$Mg(OH)_2$	29.0
Na	23.0
Na_2CO_3	53.0
$NaHCO_3$	84.0
Na_2SO_4	71.0
NaCl	58.5

Now from the above analysis, since the water contains 32 mg/L of Ca, to convert to meq/L:

$$32.1 \div 20 = 1.6 \text{ meq/L Ca}$$

$$1 \text{ meq/L Mg} = 24.31 \div 2 = 12.16 \text{ mg/L}$$
$$Mg = 7.3 \div 12.16 = 0.6 \text{ meq/L Mg}$$

$$1 \text{ meq/L Na} = 23.0 \div 1 = 23.0 \text{ mg/L}$$
$$Na = 9.2 \div 23 = 0.4 \text{ meq/L Na}$$

$$1 \text{ meq/L HCO}_3 = 61 \div 1 = 61 \text{ mg/L}$$
$$HCO_3 = 121.9 \div 61 = 2.0 \text{ meq/L HCO}_3$$

$$1 \text{ meq/L SO}_4 = 96/2 = 48 \text{ mg/L}$$
$$SO_4 = 19.2/48 = 0.4 \text{ meq/L SO}_4$$

$$1 \text{ meq/L Cl} = 35.1 \div 1 = 35.1 \text{ mg/L}$$
$$Cl = 7.1 \div 35.1 = 0.2 \text{ meq/L Cl}$$

Compiling the above calculations:

Ca^{2+}	1.6 meq/L	
Mg^{2+}	0.6 meq/L	= 2.6 meq/L
Na^+	0.4 meq/L	
HCO_3	2.0 meq/L	
SO_4^{2-}	0.4 meq/L	= 2.6 meq/L
Cl^-	0.2 meq/L	

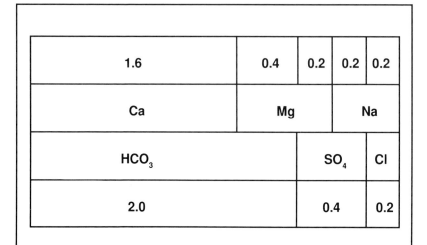

Figure 4-2. Graph of analysis.

Note that the sum of the cations is equal to the sum of the anions. The hypothetical combinations (Figure 4-2) are determined in the same manner as previously discussed.

$Ca(HCO_3)$	1.6 meq/L
$CaSO_4$	0.0 meq/L
$CaCl_2$	0.0 meq/L
$Mg(HCO_3)_2$	0.4 meq/L
$MgSO_4$	0.2 meq/L
$MgCl_2$	0.0 meq/L
$NaHCO_3$	0.0 meq/L
Na_2SO_4	0.2 meq/L
NaCl	0.2 meq/L
Total dissolved solids	= 2.6 meq/L

The concentration of each compound is determined by the application of the following formula

mg/L =
(atomic weight of cation/valence + atomic weight of anion/valence) × meq/L

mg/L $Ca(HCO_3)_2$ = (40 ÷ 2 + 61 ÷ 1) × 1.6 = 130 mg/L as $Ca(HCO_3)_2$
To convert to $CaCO_3$: 130 × 0.617 = 80.2 mg/L

mg/L $Mg(HCO_3)_2$ = (24.32 ÷ 2 + 61 ÷ 1) × 0.4 = 29.3 mg/L as $Mg(HCO_3)_2$
To convert to $CaCO_3$: 29.3 × 0.684 = 20.04 mg/L

mg/L $MgSO_4$ = (24.32 ÷ 2 + 96 ÷ 2) × 0.2 = 12.032 mg/L as $MgSO_4$
To convert to $CaCO_3$: 12.032 × 0.831 = 9.99 mg/L

mg/L Na_2SO_4 = (23 ÷ 1 + 96 ÷ 2) × 0.2 = 14.2 mg/L as $NaSO_4$
To convert to $CaCO_3$: 14.2 × 0.705 = 10.01 mg/L

mg/L NaCl = (23 ÷ 1 + 35.5 ÷ 1) × 0.02 = 11.7 mg/L as NaCl
To convert to $CaCO_3$: 11.7 × 0.856 = 10.0 mg/L

Rounding out the $CaCO_3$ equivalents, the total dissolved solids =
80 + 20 + 10 + 10 + 10 = 130 mg/L ❏

CHAPTER 5

BASIC PARAMETERS

The possible combinations of the ions having been established, it becomes necessary to discuss the methods used to measure the compounds that exist in the water. Although there are a number of parameters common to most treatment methods, perhaps the most important are those of hardness, alkalinity, pH, acidity, and salinity.

The ions that are responsible for each of the above properties are shown in Table 5-1.

It can thus be determined that calcium carbonate ($CaCO_3$) will impart both hardness and alkalinity to the water, whereas calcium chloride ($CaCl_2$) will add hardness and salinity. Magnesium bicarbonate ($Mg(HCO_3)_2$) gives hardness and alkalinity, and magnesium chloride ($MgCl_2$) will furnish hardness and salinity, and so on.

Hardness

Hardness, caused by calcium and magnesium, can be evaluated by the amount of soap required to produce a lather in water containing these ions. The greater the amount of soap required, the greater the hardness.

Acidity/Salinity

When the Cl^- and SO_4^{2-} ions combine with Mg^{2+}, Ca^{2+} of Na^+, salinity will result. When the Cl^- and SO_4^{2-} anions combine with the H^+ ion, mineral acidity is obtained.

Salinity is a characteristic that imparts a salty taste to the water. Figure 5-

Table 5-1
Hardness, Alkalinity, Acidity

Hardness	Alkalinity	Acidity &/or Salinity
Ca^{2+}	HCO_3^-	Cl^-
Mg^{2+}	CO_3^2	SO_4^{2-}
	OH^-	

Sodium (Na^+) contributes no hardness

1 indicates a special relationship between pH and alkalinity. Hence, these will be discussed jointly.

Alkalinity

An alkaline substance is one that neutralizes acid. Alkaline ions include HCO_3^{2-}, CO_3^{2-}, and OH^-.

pH

In simple terms, pH is a measure of the degree of alkalinity or acidity of water.

A water is said to be neutral if its pH is 7.0. From pH 7.0 to pH 14.0 the water is alkaline. From pH 7.0 to pH 0.0 the water is acidic.

Figure 5-1 shows pH/alkalinity relationships. The data shown in Figure 5-1 can be summarized as follows:

Bicarbonate alkalinity (HCO_3^-). When this exists alone, it exists on the pH scale only below 8.3. Since almost all natural waters have a pH between 6.5 and 7.5, all alkalinity in these natural waters will exist as bicarbonates.

Bicarbonate alkalinity + carbonate alkalinity ($HCO_3^- + CO_3^{2-}$). These alkalinities can exist together only on the pH scale between pH 8.3 and pH 9.4.

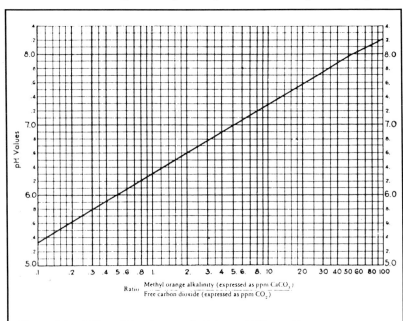

Figure 5-1a. Effect of bicarbonate alkalinity and CO_2 on pH.
Source: U.S. Filter/Permutit.

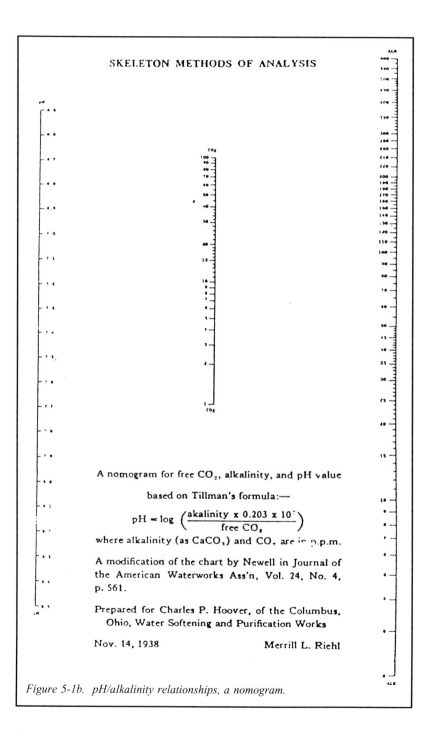

Figure 5-1b. pH/alkalinity relationships, a nomogram.

Carbonate alkalinity (CO_3^{2-}). Carbonate alkalinity can exist alone above a pH of 8.3 or in combination alkalinity in the pH range of 8.3 to 9.4. It can also exist in combination with hydroxide alkalinity (OH) above a pH of 9.4.

Carbonate alkalinity + hydroxide alkalinity ($CO_3^{2-} + OH^-$). These can exist together only at a pH above 9.4.

Caustic alkalinity (NaOH). Caustic alkalinity can exist only at a pH above 9.4. Calcium carbonate is least soluble at pH 9.4. At this pH, its solubility is 13 ppm as $CaCO_3$. Magnesium hydroxide is least soluble at pH 10.6. At this pH, the solubility is 10.2 ppm as $CaCO_3$.

While the above pH/alkalinity relationships have established the pH ranges at which various alkalinities can exist, they serve only as a guide. They do not establish the quantitative amounts of the various alkalinities. Such information is required for determination of the chemicals or treatment required, and also to determine the degree of treatment achieved.

For example, at pH of 7.0 a natural water might contain a bicarbonate alkalinity (HCO_3^-) of 50 ppm, or it might be 200 ppm. In other words, the amount of bicarbonates cannot be determined by pH alone.

To obtain these quantitative values, use is made of acid titrations (described in "Chemical Tests"). These are termed the P and M tests, so called because of the indicators that are used to determine the end points (see "Chemical Tests").

An indicator is an organic compound that changes color as the pH changes, and which is used by adding several drops of the indicator to a sample of the water to be titrated.

When P (phenolphthalein) indicator is added to a sample of water that has a pH above 8.3, the solution takes on a faint pink to deep purple color, depending on increasing pH. Below pH 8.3, the solution is colorless.

When M (methyl orange) is added, the color change is from pale yellow at pH 8.3, to orange at pH 4.3.

Table 5-2
Bicarbonate, Carbonate, Hydroxide Relationship

When	HCO_3^-	CO_3^{2-}	OH^-
P = 0	M	0	0
P = ½M	0	M	0
P = M	0	0	M
P < ½M	M-2P	2P	0
P > ½M	0	2(M-P)	2 P-M

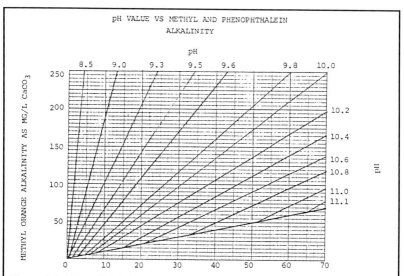

Figure 5-2. Phenolphthalein alkalinity as mg/L CaCO₃.
Source: Infilco Degremont, Inc.

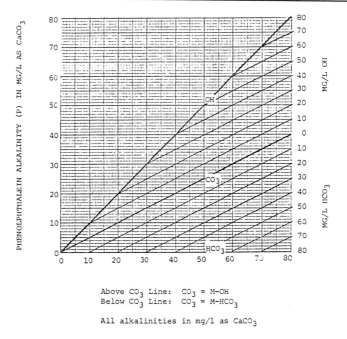

Above CO₃ Line: CO₃ = M-OH
Below CO₃ Line: CO₃ = M-HCO₃

All alkalinities in mg/l as CaCO₃

Figure 5-3. Total Alkalinity (M) in mg/L as CaCO₃.
Source: Infilco Degremont, Inc.

The **P** alkalinity test measures half of the bicarbonates ($\frac{1}{2}$ HCO$_3^-$) and all of the hydroxides (OH).

$$P = \frac{1}{2}\ HCO_3^- + OH^-$$

The **M** alkalinity test (methyl orange or methyl purple indicator) measures bicarbonates (HCO$_3^-$) plus all of the carbonates (CO$_3^{2-}$) plus all of the hydroxides (OH$^-$).

$$M = HCO_3 + CO_3 + OH$$

One word of caution: HCO$_3^-$ and OH$^-$ cannot exist in the same solution. The above data is used to achieve the criteria shown in Table 5-2.

These formulas are required to establish proper chemical balances when lime or lime/soda ash or calcium chloride are used for reduction or removal of alkalinity or hardness.❑

CHAPTER 6

STOICHIOMETRY

Stoichiometry is the study of fundamental principles: the proportion in which atoms combine, and the weight relations in a chemical reaction. Chemists have learned from the study of stoichiometric relations that the quantities of the compounds involved in a chemical reaction are proportional to their molecular weights.

Example:
$$Ca(OH)_2 + Ca(HCO_3)_2 \longrightarrow 2CaCO_3 + 2H_2O$$
$$74 + 162 = 200 + 36$$
$$236 = 236$$

This means that 74 parts of $Ca(OH)_2$ will react with 162 parts of $Ca(HCO_3)_2$ to give 200 parts of $CaCO_3$ plus 36 parts of H_2O. The figures used in this reaction are the molecular weights of each of the compounds:

Atomic weight of Ca = 40	Atomic weight of Ca = 40
Atomic weight of 20 = 32	Atomic weight of 2H = 2
Atomic weight of 2H = 2	Atomic weight of 2C = 24
___	Atomic weight of 203 = <u>96</u>

Molecular weights:
    of $Ca(OH)_2$ = 74   of $Ca(HCO_3)_2$ = 162

Example: To calculate the amount of hydrated lime (calcium hydroxide), $Ca(OH)_2$, needed to react with 120 ppm of $Ca(HCO_3)_2$:

$$
\begin{array}{ccccccc}
a & & 120 & & b & & \\
Ca(OH)_2 & + & Ca(HCO_3)_2 & = & 2CaCO_3 & + & 2H_2O \\
74 & & 162 & & 200 & & 36
\end{array}
$$

$a \div 74 = 120 \div 162$

$162a = 120 \times 74$

$a = (120 \times 74) \div 162 = 8{,}880 \div 162$

$$a = 54.8 \text{ ppm of lime } (100\%) \text{ required}$$
$$a = 54.8/0.90 = 60 \text{ ppm of lime } (90\%) \text{ required,}$$
and the dosage is $60/17.1 = 3.51$ grains/gallon (gr/gal)

Since 120 ppm = 1 lb/1,000 gal, then 60 ppm = ½ lb/1,000 gal. Likewise, the amount of $CaCO_3$ sludge that is formed is as follows:

$$b \div 200 = 120 \div 162 \text{ ppm}$$
$$162b = 120 \times 200$$
$$b = (120 \times 200) \div 162 = 24,000 \div 162$$
$$b = 148.2 \text{ ppm or}$$
$$b = 148.2 \div 120 = 1.23 \text{ lb/1,000 gal}$$

It is necessary to understand and use these relationships in order to calculate the chemicals required for the desired treatment. Further, this same knowledge will help the operator to forecast the characteristics of the treated water. There are several ways to make these determinations:

Lime Requirements
1. Determine the chemicals required for each individual component. For example, if lime is used to obtain a balanced treatment, the following calculations are required:

Lime to react with $Ca(HCO_3)_2$ and $Mg(HCO_3)_2$
+ Lime to react with Mg hardness
+ Lime to react with CO_2 if present
+ Lime to react with coagulant

Thus it is noted that some four or five stoichiometric calculations must be made.

2. A shortcut method can be used. In this method of calculation, for example, lime requirements are determined by combining the values of bicarbonates (HCO_3) and magnesium hardness, along with a predetermined factor.

Arguments can be made that the first method is more accurate, but in reality, it is the chemical control test, to be described, that makes the final determination of whether too much or too little of any chemical is being used. It would be well for purposes of this discussion to assume that either of the above calculations gives approximate value, and to let the control tests set the final chemical dosages.

Since the shortcut method is the one generally used by chemists and operators in various industries, and also to avoid confusion, only this method will be detailed.

For those who wish to pursue this subject matter, an excellent reference is the *Water and Waste Treatment Data Book* from Permutit (U.S. Filter).

For the purposes of this discussion, another analysis will be used in order to broaden the scope of this principle.

Constituent	Lab Analysis (ppm)	as ($CaCO_3$) (ppm)
Calcium (Ca)	44.88	112
Magnesium (Mg)	15.1	62
Sodium (Na)	46.5	101
Total cations		275
Bicarbonates (HCO_3)	235.0	193
Sulfates (SO_4)	10.6	11
Chlorides (Cl)	49.6	70
Total anions		274

P alkalinity = 0.0 ppm
M alkalinity = 193.0 ppm
Hardness = 112 + 62 = 174.0 ppm

Using the examples of the hypothetical combinations and pictorial representation shown in Chapter 4, "Units of Measurement and Useful Calculations," and it will be noted that the composition of this supply (in ppm as $CaCO_3$) is as follows:

Calcium bicarbonate, $Ca(HCO_3)_2$	112
Magnesium bicarbonate, $Mg(HCO_3)_2$	62
Sodium bicarbonate*, $NaHCO_3$	19
Sodium sulfate, Na_2SO_4	11
Sodium chloride, NaCl	70
Ca hardness	112
Mg hardness	62
Total hardness	174
M (total alkalinity)	193

Lime Requirements for Reduction of Total Alkalinity and Hardness
Lime (93% $Ca(OH)_2$) will react with all of the bicarbonates (201 ppm). It will also react with $MgSO_4$ and $MgCl_2$:

---

* See "$CaCl_2$ Requirements" which follows.

$$Ca(HCO_3)_2 + Ca(OH)_2 \longrightarrow \underline{2CaCO_3} + 2H_2O \ (\underline{CaCO_3} \text{ indicates a solid})$$

If $Mg(HCO_3)_2$ were present,

$$Mg(HCO_3)_2 + Ca(OH)_2 \longrightarrow MgCO_3 + \underline{CaCO_3} + 2H_2O$$

and with the addition of more lime,

$$MgCO_3 + Ca(OH)_2 \longrightarrow \underline{Mg(OH)_2} + \underline{CaCO_3}$$
$$MgSO_4 + Ca(OH)_2 \longrightarrow \underline{Mg(OH)_2} + CaSO_4$$
$$MgCl_2 + Ca(OH)_2 \longrightarrow \underline{Mg(OH)_2} + CaCl_2$$

Before the lime calculation can be made, it must be considered that the 2 gr/gal of $FeSO_4.7H_2O$, used as coagulant will react with 14 ppm of the raw water alkalinity. This then would give a revised M of 193 - 14 = 179 ppm.

Lime required = (M + Mg hardness) ÷ 100 × 0.687 = lb/1,000 gal
= (179 + 62) ÷ 100 × 0.687 = 1.65 lb/1,000 gal

It is customary to carry an OH excess of 7 ppm, which is equal to 7 ÷ 120 = 0.058 lb/1,000 gal. This then makes the total lime requirement = 1.65 + 0.0584 lb = 1.72 lb/1,000 gal.

No CO_2 was reported, and hence no lime was required for its neutralization. Had CO_2 been present, its lime requirement would be as follows:

$$ppm \ CO_2 \div 10 \times 0.156 = lb/1,000 \ gal$$

Control test. Once the P and M titrations have been made, the sufficiency of lime that furnishes the OH⁻ ion can be determined as follows (see Table 12):

2P - M = A
P = Phenolphthalein alkalinity
M = Methyl orange alkalinity
A = Excess OH

In practice, the lowest alkalinity can be obtained by carrying a small excess of hydroxide alkalinity (OH). Therefore, when A is less than (<) 2.0 ppm, add lime; when A is greater than (>) 7.0 ppm, decrease lime.

Example:

Assume P = 18 and M = 40 ppm
2 × 18 - 40 = -4, and more lime is required.

Assume P = 18 and M = 30 ppm
2 × 18 - 30 = 6,

and since this value is between 2 and 7, lime feed is sufficient.

Composition of Treated Water (Lime Treatment Only)
When calculating the composition of the treated water, calculate for the chemicals added by treatment. For example:

1. Two gr/gal of $FeSO_4.7H_2O$ used as a coagulant will add 14 ppm of SO_4, which will remain in the water. Calculate the total SO_4 as follows:

SO_4 in raw water	11 ppm
SO_4 added	14 ppm
Total SO_4 in treated water	25 ppm

2. The addition of 12 ppm of chlorine, whether it be as Cl_2 gas or NaOCl, will add 12 ppm × 1.41 = 16.92 (as $CaCO_3$). Round it off to 17 ppm. Calculate the total chlorides as follows:

Chlorides in raw water	70 ppm
Chlorides added	17 ppm
Total chlorides in treated water	87 ppm

If NaOCl is used, 17 ppm of Na ($CaCO_3$) will be added to bring the total sodium to 101 + 17 = 118 ppm.

3. The chemical reactions shown previously indicate that all of the calcium carbonate and magnesium hydroxide have been precipitated. Assume that 15 ppm of $CaCO_3$ are theoretically soluble as are some 8 ppm of $Mg(OH)_2$. These will remain in solution in the treated water.

Since both of these compounds will impart alkalinity to the treated water, calculate as follows:*

Ca alkalinity	15 ppm
Mg alkalinity	8 ppm
Na alkalinity	19 ppm
Alkalinity caused by excess lime	7 ppm
M = Total alkalinity	49 ppm

4. To recapitulate the foregoing, the treated water will have the following composition, all in ppm as $CaCO_3$:

Calcium Hardness		Sum
As soluble $CaCO_3$	15	
As excess $Ca(OH)_2$	7	
As SO_4 from $FeSO_4$	14	
	36	36

* H and M values may be low, but are used for the sake of simplicity.

Magnesium (Mg Hardness)		*Sum*
As soluble $Mg(OH)_2$	8	8

Sodium (Na)		
As original	101	
From NaOCl	17	
	118	118
Total Cations		162

M (Alkalinity)		
As soluble $CaCO_3$	15	
As soluble $Mg(OH)_2$	8	
As excess $Ca(OH)_2$	7	
As $NaHCO_3$	19	
	49	49

Sulfates (SO_4)		
Original	11	
From $FeSO_4$	14	
	25	25

Chlorides (Cl)		
Original	70	
From NaOCl	17	
	87	87
Total Anions		161

$$2P - M = 7$$
$$2P - 49 = 7$$
$$2P = 56$$
$$P = 28$$

All of these calculations show theoretical amounts. It should be noted that an excess of lime will be determined by the P and M tests. In actual practice, the above are very close to the actual requirements and should be considered the starting dosages.

Calcium Chloride Requirements

Calcium chloride (75% $CaCl_2$ solution) is employed for the removal of sodium bicarbonate alkalinity. When calculating the amount of lime ($Ca(OH)_2$) required for hardness and total alkalinity reduction, it was assumed that any coagulant ($FeSO_4$) would react with the $Ca(HCO_3)_2$ and/or $Mg(HCO_3)_2$, leaving the sodium bicarbonate ($NaHCO_3$) in its original quantity.

The $Ca(OH)_2$ will not react with $NaHCO_3$, and $CaCl_2$ (or $CaSO_4$) is used to remove this alkalinity.

$$2NaHCO_3 + CaCl_2 \longrightarrow Ca(HCO_3)_2 + 2NaCl$$

and

$$Ca(HCO_3)_2 + Ca(OH)_2 \longrightarrow \underline{2CaCO_3} + 2H_2O,$$

and it can be assumed that the original lime dosage was sufficient to react with the above $Ca(HCO_3)_2$.

Therefore, the quantity of $CaCl_2$ required for removal of the $NaHCO_3$ is calculated as follows:

$$(M - H) \div 100 \times 1.23 = \text{lb of } 75\% \text{ CaCl}_2/1{,}000 \text{ gal}$$
$$(193 - 174) \div 100 \times 1.23 = 0.23 \text{ lb of } 75\% \text{ CaCl}_2/1{,}000 \text{ gal}$$

There has been some question concerning the use of 75% $CaCl_2$ for ingested products, and in such instances it is required to use food-grade $CaCl_2$. In this situation,

$$(M - H) \div 100 \times 1.11 = \text{lb of CaCl}_2/1{,}000 \text{ gal}$$
$$(193 - 174) \div 100 \times 1.1 = 0.21 \text{ lb of CaCl}_2/1{,}000 \text{ gal}$$

The difference in dosage does not seem too great, but food-grade $CaCl_2$ is considerably more expensive than the non-food-grades, and when large quantities of water are treated, the difference in cost is appreciable.

Control Test for Calcium Chloride Requirements
In the formula,

$$M - \text{hardness} = \text{excess CaCl}_2$$
$$\text{Methyl orange alkalinity (M)} - \text{hardness} = \text{excess CaCl}_2$$

Use enough $CaCl_2$ so that M - hardness = 7.0 ppm.

To check the lime feed, it is necessary to apply the test for lime previously discussed: $2P - M = A$. Therefore to have a balanced treatment when using both $Ca(OH)_2$ and $CaCl_2$ or $CaSO_4$,

$$A = 2P - M = 2.0 - 7.0 \text{ ppm}$$

and

$$M - \text{hardness} = \text{excess CaCl}_2 = 7.0 \text{ ppm}$$

Composition of Treated Water
When calculating the composition of the treated water (lime and calcium chloride treatment), consideration must be given to the following facts:

1. The use of ferrous sulfate ($FeSO_4 \cdot 7H_2O$) as a coagulant will add SO_4 (as $CaCO_3$). Since the usual dosage of $Fe_2SO_4 \cdot 7H_2O = 2$ gr/gal $= 34.2$ ppm (round to 35 ppm), then the SO_4 will be increased by $35/20 \times 8 = 14$ ppm SO_4 as $CaCO_3$. Therefore the total SO_4 will be as follows:

SO_4 in raw water	= 11 ppm
+ SO_4 added	= 14 ppm
Total SO_4	= 25 ppm

2. The addition of 12 ppm of chlorine, whether as Cl_2 gas or as $NaOCl$, will add chlorides to the treated water. This will be in the amount of 12 ppm $\times 1.41 = 16.92$ ppm as $CaCO_3$ (round to 17 ppm).

At the same time, the use of 0.21 lb/1,000 gal of $CaCl_2$ will increase the chlorides by the following: 0.21 lb/1,000 gal $\times$ 120 ppm $= 25.2$ ppm. Therefore, the total chlorides in the treated water will be as follows:

Chlorides in raw water	= 70 ppm
Chlorides added by chlorine	= 17 ppm
Chlorides added by $CaCl_2$	= 25 ppm
Total chlorides in treated H_2O	= 112 ppm

3. The chemical reactions shown for the use of lime and calcium chloride indicate that all of the calcium carbonate and magnesium hydroxide have been precipitated. However, assume that 15 ppm of $CaCO_3$ are theoretically soluble, as are some 8 ppm of magnesium hydroxide. These will remain in solution in the treated water.

Since both carbonates and hydroxides will impart alkalinity, the total alkalinity of the treated water will be as follows:

M from $CaCO_3$	= 15 ppm
M from $Mg(OH)_2$	= 8 ppm
	23 ppm

Also, remember that an excess of 7 ppm of $Ca(OH)_2$ was added. This in addition to the above will give a final alkalinity of $23 + 7 = 30$ ppm. Since both calcium and magnesium impart hardness, the hardness of the treated water will be as follows:

H from $CaCO_3$	= 15 ppm
H from $Mg(OH)_2$	= 8 ppm
H from $FeSO_4$ reaction	= 14 ppm
H from excess lime	= 7 ppm
	44 ppm

Similarly, 7 ppm of CaH will be added because of the excess of lime. Therefore, the total hardness of the treated water will be 23 + 7 = 30 ppm.

4. To recapitulate the foregoing, the treated water will have the following composition:

Calcium hardness (Ca)	= 37 ppm
Magnesium hardness (Mg)	= 8 ppm
*Sodium (Na)	= 123 ppm
Total cations	= 168 ppm
Alkalinity (M)	= 30 ppm
Sulfates (SO_4)	= 25 ppm
Chlorides (Cl)	= 112 ppm
Total anions	= 168 ppm

All of the calculations show theoretical amounts. It should be noted that an excess of both lime and calcium chloride will be determined by the P, M, and hardness tests in actual practice. The above are very close to the actual requirements and should be considered as the starting dosages.

It is noted that in the above calculations, sufficient chemicals (lime and chloride) are used to capture all of the bicarbonate alkalinity and hardness.

Lime / Soda Ash Requirements
For the removal of noncarbonate hardness, 98% soda ash (Na_2CO_3) can be used. Noncarbonate hardness occurs when hardness is greater than total alkalinity. In the previous calculations for required chemicals, it was purposely arranged that all of the hardness in the water was comprised of $Ca(HCO_3)_2$ and $Mg(HCO_3)_2$. This was done for the sake of simplicity.

As a result, it was possible to reduce all of the hardness to a minimum using lime, as follows:

$$Ca(HCO_3)_2 + Ca(OH)_2 \longrightarrow \underline{2CaCO_3} + 2H_2O$$
$$Mg(HCO_3)_2 + Ca(OH)_2 \longrightarrow MgCO_3 + \underline{CaCO_3} + 2H_2O$$
$$\text{and } MgCO_3 + Ca(OH)_2 \longrightarrow Mg(OH)_2 + \underline{CaCO_3}$$

However, had $CaSO_4$, $CaCl_2$, $MgSO_4$, or $MgCl_2$ been present, these noncarbonates would have reacted with lime as follows:

$$CaSO_4 + Ca(OH)_2 \longrightarrow \text{ No reaction}$$

* Total sodium arrived at by subtracting the Ca + Mg (37 + 8) = 45 ppm from the total anions (168 - 45) = 123.

$$CaCl_2 + Ca(OH)_2 \longrightarrow \text{No reaction}$$
$$MgSO_4 + Ca(OH)_2 \longrightarrow \underline{Mg(OH)_2} + CaSO_4$$
$$MgCl_2 + Ca(OH)_2 \longrightarrow \underline{Mg(OH)_2} + CaCl_2$$

Note that the magnesium hardness was removed as a precipitate, but in each reaction $CaSO_4$ and $CaCl_2$ (both hardness-containing compounds) remained in solution. If minimum hardness is desired, then it becomes necessary to use soda ash (Na_2CO_3). The compounds will react as follows:

$$CaSO_4 + Na_2CO_3 \longrightarrow \underline{CaCO_3} + Na_2SO_4$$
$$CaCl_2 + Na_2CO_3 \longrightarrow \underline{CaCO_3} + 2NaCl$$

The same would hold true of magnesium noncarbonate solids.

$$\text{Soda ash} = (\text{hardness - M}) \div 100 \times 0.9 \text{ lb/1,000 gal}$$

Example: If a water has a total hardness of 150 ppm and an alkalinity of 100, the Ca + Mg alkalinity = 100, and noncarbonate hardness = 150 - 100 = 50.

Example: To reduce the alkalinity and hardness to approximately 35 ppm, it is necessary to use:

$$\text{Lime for Ca + Mg alkalinity} = M \div 100 = 100 \div 100 \times 0.687$$
$$= 1 \times 0.687 = 0.687 \text{ lb/1,000 gal}$$

Plus soda ash for noncarbonate hardness =
$$(\text{hardness - M}) \div 100 \times 0.9 \div 1,000 \text{ gal}$$
$$= 50 \div 100 \times 0.9 \text{ gal}$$
$$= 0.45 \text{ lb/1,000 gal}$$

Control Test for Lime/Soda Ash Requirements
In order that the alkalinity reduction is kept at maximum, the lowest hardness obtainable will be limited as follows:

$$2 P - M = (\text{excess hydrate}) = A$$
$$(M - P) = B \text{ (normal carbonate)}$$
$$B - \text{hardness} = C \text{ (excess soda ash)}$$

(a) If A is negative, further calculations are of no value. First get the A value to between 2 and 7 ppm.

(b) Next add enough soda ash so that B - hardness = 0

(c) If A is positive and B is negative, more soda ash is required. If A is positive and B is positive, be certain that the B value is as near zero as it is possible to attain. Too large an excess of soda ash will impart a caustic taste.

It is left to the reader to composite an analysis of treated water using the methods discussed.❑

CHAPTER 7
SOLIDS PRODUCED BY CHEMICAL REACTIONS

It has been shown that lime will react with $Ca(HCO_3)_2$, $Mg(HCO_3)_2$, $MgSO_4$, and $MgCl_2$ to reduce hardness and bicarbonate alkalinity. These reactions produce insoluble solids. Lime will not react with $CaSO_4$ or $CaCl_2$. If the hardness contributed by these latter compounds must be removed, it is necessary to use soda ash (Na_2CO_3).

Lime will not react with sodium bicarbonate, and to remove alkalinity it is necessary to use calcium chloride. This plus added lime will produce solids (calcium carbonate).

Common to each of these treatments is the use of a coagulant to effect a better separation of these solids from the water in the reactor units. These coagulants — $Al_2(SO_4)_3$ (aluminum sulfate); or an iron salt such as $FeSO_4$ (ferrous sulfate), $Fe_2(SO_4)_3$ (ferric sulfate), or $FeCl_3$ (ferric chloride) — will also produce solids.

If a complete analysis is not available, an approximation of the solids produced can be determined as follows:

M_r = Raw water alkalinity (in ppm)
M_t = Treated water alkalinity (in ppm)
T_r = Raw water color + turbidity (in ppm)
T_t = Treated water color + turbidity (in ppm)

(a) Solids produced from lime treatment = $(M_r \times 2 - M_t) \times gpm \times 0.0005$ pounds per hour (lb/h)

(b) Solids produced from $FeSO_4$ = gr/gal $FeSO_4 \times gpm \times 0.003$ lb/h

(c) Solids produced from $Al_2(SO_4)_3$ = gr/gal $Al_2(SO_4)_3 \times gpm \times 0.002$ lb/h

(d) Solids produced from turbidity = $(T_r - T_t) \times gpm \times 0.0005$ lb/h

Total solids produced = sum of above

Note: Use either (b) or (c), not both.

Solids Produced from Lime Treatment

If an analysis is available, and the compounds have been determined through the use of hypothetical combinations, the amount of solids produced from lime treatment can be calculated as follows:

Solids (in ppm) caused by the use of coagulants:

ppm of color	$\times 1 =$ _____ ppm
Decrease in turbidity	$\times 1 =$ _____ ppm
gr/gal $Al_2(SO_4)_3$	$\times 4.5 =$ _____ ppm
gr/gal $FeSO_4$	$\times 6.6 =$ _____ ppm
Total solids	$=$ _____ ppm

Since only one of the coagulants is used, do not include both the $Al_2(SO_4)_3$ and $FeSO_4$. The solids produced from these reactions would be as follows:
ppm of total solids $\times$ gpm $\times$ 0.0005 lb/h

The solids (in ppm) produced by the reaction of lime are calculated as follows (all compounds must be as $CaCO_3$):

ppm of $Ca(HCO_3)_2$	$\times 2.0 =$ _____ ppm
ppm of $Mg(HCO_3)_2$	$\times 1.88 =$ _____ ppm
ppm of $MgSO_4$	$\times 0.48 =$ _____ ppm
ppm of $MgCl_2$	$\times 0.61 =$ _____ ppm
Total solids	$=$ _____ ppm

Since some of the above solids, namely calcium and magnesium, are soluble to the extent of the final alkalinity (M)*, then this final alkalinity must be subtracted from the above value:

$$(\text{ppm solids - ppm M as above}) = \text{net solids}$$

Solids from lime produced = net solids $\times$ gpm $\times$ 0.0005 lb/h. Then total solids = solids from coagulant plus solids from lime treatment.

Solids Produced from Lime / Soda Ash Treatment

Using values as $CaCO_3$, the solids produced from lime / soda ash treatment can be determined as follows:

ppm of $Ca(HCO_3)_2$	$\times 2.0 =$ _____ ppm
ppm of $Mg(HCO_3)_2$	$\times 1.88 =$ _____ ppm
ppm of $CaSO_4$	$\times 0.735 =$ _____ ppm
ppm of $CaCl_2$	$\times 0.91 =$ _____ ppm
ppm of $MgSO_4$	$\times 0.32 =$ _____ ppm
ppm of $MgCl_2$	$\times 1.66 =$ _____ ppm
Total solids	$=$ _____ ppm

*Approximately 30 ppm (includes 7 ppm of excess lime)

$$\text{Total solids produced} = \text{net solids} \times \text{gpm} \times 0.0005 \text{ lb/h}$$
$$\text{Net solids} = \text{total solids} - 35 \pm \text{ppm}$$

To this value of net solids, add the solids produced from use of coagulant, as explained above under "Solids Produced from Lime Treatment."

Determination of Solids Using P, M, and Hardness[*]

Table 7-1 is based on water analyses that report alkalinity (M), hardness, and magnesium hardness in ppm as calcium carbonate. If the hardness is reported as calcium (Ca) and/or magnesium (Mg), multiply these values by the following factors to convert them to ppm of hardness in terms of calcium carbonate:

$$\text{ppm of Ca} \times 2.5 = \text{ppm of Ca hardness as calcium carbonate}$$
$$\text{ppm Mg} \times 4.11 = \text{ppm of Mg hardness as calcium carbonate}$$

After obtaining the values of the calcium and magnesium hardness as calcium carbonate, add these values together to obtain the total hardness as calcium carbonate.

If values are reported in grains per gallon, multiply gr/gal by 17.12 to obtain ppm. (See Table 7-1.) To find the pounds of solids formed per hour, insert the total ppm of solids formed by treatment, as determined in Table 7-1, into the following formula:

$$\text{ppm of solids formed} \times \text{gpm water flow} \times 0.0005 = \text{lb of solids formed /h}$$

Clarification treatment. If the treating plant is used for clarification only, use Table 7-1a. From the water analysis, enter the ppm of color and ppm of turbidity. Enter the gr/gal of the coagulant employed (such as aluminum sulfate or ferrous sulfate). Multiply these values by the indicated factors and enter the products in the right-hand column. Add the products together to find the total ppm of solids formed.

Softening treatment with lime. When the water does not contain sodium carbonate (hardness is greater than M), or when the water does contain sodium carbonate (M is greater than hardness), and calcium chloride is used to react with the sodium carbonate, fill in the appropriate values in Table 7-1a and Table 7-1b, complete the multiplication, and total the products. Carry the total out to the right of the table.

When the water contains sodium carbonate (M is greater than hardness) but calcium chloride is not used to react with sodium carbonate, fill in the appropriate values in Table 7-1a and 7-1c, complete the multiplication, and total the products.

[*]Courtesy, Infilco Degremont, Inc.

Carry the totals out to the right of the tables.

Add the total solids formed from Table 7-1a to the total from Table 7-1b and 7-1c to find the total solids produced with lime treatment.

Softening: complete lime / soda ash treatment. When both lime and soda ash are used for softening treatment, calculate the total solids produced with lime treatment as directed above and add this total to the remainder found in Table 7-1d to find the total solids produced with lime / soda ash treatment.

Now that the weight of solids formed per hour has been calculated, it is known that this is the amount of solids that must be discharged each hour to maintain

---

Table 7-1
Calculation of Total Solids by Lime / Soda Ash Treatment

Color	_____ *ppm*	$\times 1.0 =$	_____ *ppm*
Turbidity* $(T_r\text{-}T_t)$	_____ ppm	$\times 1.0 =$	_____ ppm
Aluminum sulfate	_____ gr/gal	$\times 4.5 =$	_____ ppm
Ferrous sulfate	_____ gr/gal	$\times 6.6 =$	_____ gr/gal
Ferric chloride	_____ gr/gal	$\times 11.3 =$	_____ gr/gal
Ferric sulfate	_____ gr/gal	$\times 8.0 =$	_____ gr/gal
Activated silica	_____ ppm	$\times 1.0 =$	_____ ppm

(1a) Total solids formed $=$ _____ ppm

T_r = Raw water turbidity, T_t = treated water turbidity

M	_____ ppm	$\times 2.0 =$	_____ ppm
Mg hardness	_____ ppm	$\times 0.58 =$	_____ ppm

(1b) Total solids formed $=$ _____ ppm

_____ ppm M - hardness	_____ ppm	$\times 1.0 =$	_____ ppm
M	_____ ppm	$\times 2.0 =$	_____ ppm
Mg hardness	_____ ppm	$\times 0.58 =$	_____ ppm

(1c) Total solids formed $=$ _____ ppm

_____ ppm hardness minus_____ ppm M $\times 1.0 =$ _____ ppm

(1d) Total solids formed by treatment $=$ _____ ppm

Courtesy, Infilco Degremont, Inc.

equilibrium. To calculate the required rate of sludge discharge, the weight of solids per gallon or per liter of sludge must now be determined.

Weight of sludge. Weigh an empty bottle, preferably of the desired unit-of-volume capacity (1 gallon or 1 liter). Fill the bottle with sludge and weigh it. Subtract the weight of the empty bottle from the full weight to determine the weight of the sludge.

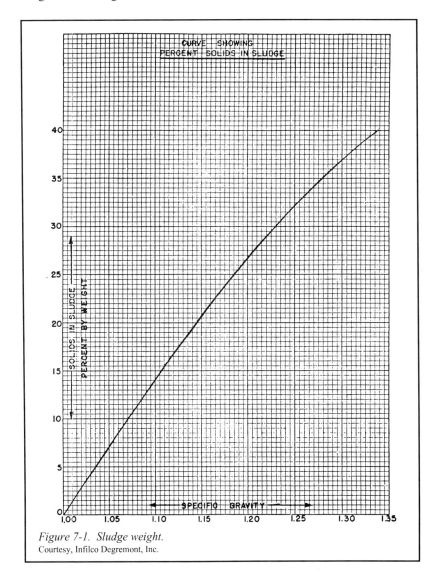

Figure 7-1. Sludge weight.
Courtesy, Infilco Degremont, Inc.

Specific gravity of sludge. Fill the same bottle with filtered raw water and weigh it. Subtract the weight of the empty bottle to determine the weight of the water. Divide the weight of the sludge by the weight of the equal amount of water to determine the specific gravity of the sludge.

Percent by weight of solids in the sludge. At the bottom of the graph in Figure 7-1, find the specific gravity just determined. Follow the vertical line upward to the diagonal line. Follow the horizontal line, intersected at the diagonal line, to the left and read the percent by weight of solids in the sludge.

Weight of solids per unit of volume (gallons or liter) of sludge. Use one of the following formulas to find the weight per unit of volume in the sludge:

Pounds per gallon =
    (8.337 × % by weight of solids × specific gravity of sludge)/100

Grams per liter =
    10 × % by weight of solids × specific gravity of sludge

Volume to be discharged per hour. Find the rate of sludge discharge in gallons per hour or liters per hour with the graph in Figure 7-1.

Example:
Assume 520 lb of solids are formed per hour, the weight of the water in a 1-gal bottle is 8.337 lb, and the weight of the same amount of sludge is 9 lb.

Specific gravity of sludge = 9 ÷ 8.337 = 1.08

Percent by weight of solids in sludge =
    11.5% (from line for specific gravity) (Figure 7-1).

Weight of solids per unit of volume =
    (8.337 × 11.5 × 1.08) ÷ 100 = 1.04 lb of solids per gallon of sludge.

Rate of sludge discharge = 520 ÷ 1.04 = 500 gal/h ❑

CHAPTER 8

COAGULATION, FLOCCULATION, AND SEDIMENTATION

Clarification, as practiced in the treatment for product water, can be described as the removal of solid particles. These are particles that are present in the water as it enters the manufacturing plant, and/or particles that come from chemical compounds added for treatment purposes.

These particles can be colloidal in size (organic matter), or comparatively large particles of substances such as calcium carbonate that are usually products of reaction. The compounds that aid in the removal of these particles are called coagulants, and the principle of clarification embodies several unit processes.

Coagulation is the formation of insoluble jellylike particles through the reaction of a coagulant with the alkalinity and hardness that exists in the water. If such do not exist, they are added along with the coagulant.

Flocculation is the collection and agglomeration of these jellylike particles so that they will result into a mass that will enmesh smaller particles. This activity will result in larger and denser material that will settle out from the water at a greater velocity.

Sedimentation is the settling out of the flocculated particles so they can be separated from the water.

Prechlorination is the addition of a chlorine-containing compound (chlorine gas or hypochlorite) in order to destroy any organic matter that may be present. As a rule the presence of organic matter is detrimental to good coagulation.

Aluminum sulfate, $Al_2(SO_4)_3 \cdot 14H_2O$, which is also known as filter alum, or commercial alum, and certain iron salts are among the most popular coagulants. Other aluminum salts in the form of either ammonia alum, $Al_2(SO_4)_3 \cdot (NH_4)_2SO_4 \cdot 24H_2O$, or potash alum, $Al_2(SO_4)_3 \cdot K_2SO_4 \cdot 24H_2O$, are also used widely in municipal or large industrial plants because of cost and availability. However; as will be noted later, some of these salts may have some limitation when compared to the iron coagulants.

In all cases, the acidic coagulants will react with alkalinity. If raw water does not contain sufficient alkalinity, then alkalinity must be added to complete the desired reaction.

At any rate, the following are the coagulants most likely to be used:

Filter alum	Aluminum sulfate	$Al_2(SO_4)_3 \cdot 14H_2O$
Soda alum	Sodium aluminate	$Na_2Al_2O_3$
Copperas	Ferrous sulfate	$FeSO_4 \cdot 7H_2O$
Ferri-chlor	Ferric chloride	$FeCl_3$
Ferrifloc	Ferric sulfate	$Fe_2(SO_4)_3 \cdot 9H_2O$

Aluminum Sulfate

Although aluminum sulfate, $Al_2(SO_4)_3 \cdot 14H_2O$, is by far the most widely used coagulant, it is not without some limitations where used for product water. It functions best in the narrow pH range between 5.8 and 7; therefore, it would not be wise to use alum where alkalinity reduction is practiced, since the treated water in this case will be pH 9.5 or even greater.

If dissolved alumina, Al_2O_3, is detrimental to the product water, extreme care must be taken since improper dosages can result as follows:

$$Al_2(SO_4)_3 + 3H_2O \text{ (Alumina)} ----> Al_2O_3 + 3H_2SO_4$$

This can occur in beverage water, for example, where the addition of carbon dioxide to a water containing alumina can result in the formation of precipitates in the finished beverage. If such does occur, or if the plant supply is of a variable nature, new jar tests are required, at which time tests should be made on the treated sample to assure the absence of alumina.

If the supply of water to the plant is of a chemically constant quality, there is no reason to suspect that the original dosages will not serve the purpose, but caution is advised. The following two reactions take place if the plant supply contains sufficient bicarbonates (HCO_3) to react with the alum that is added:

$$Al_2(SO_4)_3 + 3Ca(HCO_3)_2 ----> \underline{2Al(OH)_3} + 6CO_2 + 3CaSO_4$$
$$Al_2(SO_4)_3 + 3Mg(HCO_3)_2 ----> \underline{2Al(OH)_3} + 6CO_2 + 3MgSO_4$$

Note: Approximately 2 gr/gal will be required for proper coagulation. This dosage will require at least 16 ppm of alkalinity (HCO_3).

If alkalinity must be added to the plant supply, the following reactions take place:

$$Al_2(SO_4)_3 + 3Ca(OH)_2 \text{ (lime)} ----> \underline{2Al(OH)_3} + 3CaSO_4$$
(2 gr/gal of alum will require ±0.775 gr/gal of lime)

$$Al_2(SO_4)_3 + 3Na_2CO_3 \text{ (soda ash)} + 3H_2O ----> \underline{2Al(OH)_3} + 3Na_2SO_4 + 3CO_2$$
(2 gr/gal of alum will require ±1.0 gr/gal of soda ash)

$$Al_2(SO_4)_3 + 6NaOH \text{ (Caustic)} ----> \underline{2Al(OH)_3} + 3Na_2(SO_4)$$
(2 gr/gal of alum will require 0.76 gr/gal of caustic)

Of course, only one of the alkalinity-adding compounds will be used at any one time.

A further word of caution: Alum is also sold as *ammonia alum* $Al_2(SO_4)_3 \cdot (NH_4)_2SO_4 \cdot 24H_2O$; or *potash alum* $Al_2(SO_4)_3 \cdot K_2SO_4 \cdot 24H_2O$. If one of these must be used, *avoid* use of ammona alum, since the addition of chlorine into the treatment process will result in the formation of a chloramine that could prove to be a problem later in the process.

Iron Coagulants: Copperas, "Ferrichlor," and "Ferrifloc"
The optimum pH range for iron coagulants is 7.7 and higher. As a result, these are used when lime is required for hardness or alkalinity reduction, or when it is desired to form floc particles to enmesh turbidity, color, or organic matter from the plant supply.

Copperas. Each of the above coagulants has its advantages and disadvantages. The use of $FeSO_4$ (ferrous sulfate, or copperas) in combination with lime and chlorine seems to have become the most popular. As a matter of fact, this combination has become referred to as the "universal dosage" for product water treatment. When referred to in these terms, it means an $FeSO_4$ dosage of 2 gr/gal, sufficient lime to carry a slight OH^- excess, and a chlorine dosage to impart a chlorine residual of 6 to 8 ppm.

This dosage eliminates the possibility of dissolved iron in the final water, in contrast to the possible alumina problem that might be encountered when using aluminum sulfate.

If bicarbonate alkalinity is present in the raw water:

$$FeSO_4 + Ca(HCO_3)_2 ----> \underline{Fe(OH)_2} + CaSO_4 + 2CO2$$
$$2FeSO_4 + Cl_2 + 3Ca(HCO_3)_2 ----> \underline{2Fe(OH)_3} + 2CaSO_4 + CaCl_2 + 6CO_2$$

(Approximately 2 gr/gal of ferrous sulfate are required for proper coagulation. This dosage will require ±1.2 gr/gal of lime.) If not,

$$FeSO_4 + Ca(OH)_2 ----> \underline{Fe(OH)_2} + CaSO_4$$
(Two gr/gal of $FeSO_4$ will require ±0.9 gr/gal of lime.)

$$FeSO_4 + Na_2CO_3 ----> \underline{FeCO_3} + Na_2SO_4$$
(Two gr/gal of $FeSO_4$ will require ±1.2 gr/gal of soda ash.)

Note that the end product in some cases is $Fe(OH)_2$, which is a greenish-brown solid. In actual practice, any oxygen present in the water will give the following:

$$4Fe(OH)_2 + O_2 + 2H_2O ----> 4Fe(OH)_3$$
$$(Fe(OH)_3 \text{ is the brown floc that is preferred.})$$

Ferric sulfate. Sold as Ferrifloc, $Fe_2(SO_4)_3$ needs a pH of 4 to 11. If sufficient alkalinity exists in water:

$$Fe_2(SO_4)_3 + 3Ca(HCO_3)_2 ----> \underline{2Fe(OH)_3} + 3CaSO_4 + 6CO_2$$

Assume that the dosage for good coagulation is 2 gr/gal, then 1.5 gr/gal of bicarbonates are required. If lime is added for alkalinity reduction or for softening (with required quantity of lime determined by the alkalinity and/or hardness of the plant supply), the following reaction takes place:

$$Fe_2(SO_4)_3 + 3Ca(OH)_2 ----> \underline{2Fe(OH)_3} + 3CaSO_4$$

The ferric hydroxide $(Fe(OH)_3)$ in either case is a brown color, and it has good settling characteristics.

Ferric chloride. Sold as Ferrichlor, $FeCl_3$ needs a pH of 4 to 11. As in the above cases, if sufficient alkalinity is present in supply water:

$$2FeCl_3 + 3Ca(HCO_3)_2 ----> \underline{2Fe(OH)_3} + 3CaCl_2 + 6CO_2$$

If lime is used in treatment:

$$2FeCl_3 + 3Ca(OH)_2 ----> \underline{2Fe(OH)_3} + 3CaCl_2$$

The dose used for coagulation is approximately 2 gr/gal of $FeCl_3$.

Jar Tests
In those examples for which coagulation and settling have been discussed, it has been assumed that the chemical dosages required would be as follows:

Aluminum or ferrous sulfate - 2 gr/gal.

Chlorine - Sufficient to maintain an 8-ppm residual.

Lime - Sufficient to reduce the alkalinity to the desired parameter, or to react with the ferrous sulfate or other coagulant if alkalinity reduction is not required.

Further, the dosages and base charges of these chemicals are always furnished to the operator by the manufacturer or quality control, both of whom have previously pretested the raw water. There may be occasions, however, when because of changing raw water characteristics different chemical dosages are required. These new dosages will be determined through a procedure known as *jar tests*. This consists of adding known dosages of chemicals to a known volume of water, stirring, flocculating, allowing to settle, and selecting the apparent best results. The test proceeds as follows:

a) If alkalinity reduction is necessary, the proper amounts of lime and/or calcium chloride required for jar tests are calculated. Since they will have a positive effect upon coagulation, these chemicals should be used. Further, in case of alkalinity reduction, the P and M guidelines test must be adhered to. A pH range of from 8.3 to 10.4 will be the rule.

b) If the raw water is at or below the required alkalinity, the tests will be used to determine the amount of coagulant and perhaps the small amount of lime and/or soda ash required to establish the optimum pH for best coagulation. Since there can be no P alkalinity below pH 8.3, and since there is a good possibility that best coagulation and settling can be obtained at a pH value below 8.3, the P and M guidelines will not necessarily hold true for raw water with low alkalinity.

c) In all cases the calculated amount of chlorine should be added to the raw water first, as prechlorination of raw water will always aid in the coagulation. The chlorine dosage used in actual plant operations is about 12 ppm (0.7 gr/gal).

d) To determine the proper dosages, it is necessary to prepare stock solutions of the chemicals that will be used in the determinations. Also it will be necessary to make concurrent tests in a number of 1-liter (1,000-cm^3) beakers. After the various dosages are added to the different beakers, it is necessary to rapidly mix for 1 minute and then to slowly stir the samples for from 5 to 10 minutes until the reaction is complete. While the samples are being stirred, they should be closely observed to note whether or not a good floc is being formed, and a notation is made.

Following the stirring, the samples are allowed to settle. A good floc will settle in 2 to 5 minutes. The less time taken, the better the floc. Another notation is then made.

The supernatant is observed and a record is made of its quality, whether it is clear or hazy. It will be only after such a series of determinations that the operator can then judge which is the best combination of chemicals to use.

Stock Solutions

The procedure is as follows:

a) Make a stock solution of chemicals by dissolving 17.1 grams of the chemicals in 1,000 cm^3 of distilled water. Assuming that ferrous sulfate and lime are to be tested, the operator will dissolve 17.1 g of FeSO$_4$ in 1,000 cm^3 of water. He or she will also dissolve 17.1 g of Ca(OH)$_2$ (calcium hydroxide) in 1,000 cm^3 of water to give the second stock solution. *Each cm^3 of these solutions, when added to 1,000 cm^3 of water, will give a dosage of 1 gr/gal.*

To make up the chlorine stock solution, dissolve 20 cm^3 of new 5.25% hypochlorite solution (Clorox) in 1,000 cm^3 of water. Each cm^3 of this stock solution added to 1,000 cm^3 of raw water will give a dosage of 1 ppm. It is suggested to use a starting dosage of 12 cm^3/1,000 cm^3, which will usually give a residual of 8 ppm.

b) These samples can be stirred by hand, but if much of this testing is to be performed, the operator should make use of an automatic "gangmixer" (see Figure 8-1). This is electrically driven, and it is able to stir some 6 samples at the same time. The mixture of chemicals and water should be rapidly mixed for 1 minute, then slowly stirred for 5 to 10 minutes.

Using 1,000 cm^3 samples and adding the specific quantities, and assuming that tests using ferrous sulfate and soda ash are used, the chemicals are varied in amount at the discretion of the operator. Table 8-1 shows a typical log sheet.

It would be obvious that sample #6 would result in the best dosage of chemicals, and this then would be used to calculate the base charges to be used

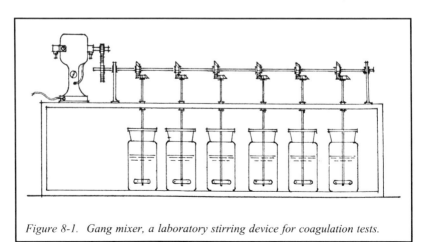

Figure 8-1. Gang mixer, a laboratory stirring device for coagulation tests.

with the chemical feeders.

It is quite possible that none of the samples would produce good results. In that case, the testing would have to be repeated using different amounts of $FeSO_4$ and $Ca(OH)_2$, or perhaps different chemicals.

When using $FeSO_4$, $Ca(OH)_2$, and Cl_2, the chlorine is fed at a dosage of 12 ppm; the $FeSO_4$ dosage is set at 2 gr/gal; and sufficient lime is used so that 2 P-M = 2 to 7 ppm. There may be times when the $FeSO_4$ dosage should be increased to, for example, 3 gr/gal, and the lime dosage increased accordingly until 2 P-M = 2 to 7 ppm.

There are other criteria that can and will affect the efficiency of coagulation and flocculation:

Sequestering Agents
In order to avoid scaling or to control corrosion in distribution lines, municipalities may at times resort to the addition of a sodium polyphosphate to the water supply. This will cause sequestering or chelating of those troublesome substances, and this may sometimes affect coagulation. To overcome such problems, it is necessary to increase the coagulant, thus lowering the pH to break down the chelating agent.

The use of $CaCl_2$, calcium chloride, will react to precipitate the phosphate as follows:

$$2Na_3PO_4 + 3CaCl_2 ----> \underline{Ca_3(PO_4)_2} \text{ (precip.)} + 6NaCl$$

Water Temperature
When the temperature of the raw water starts to approach about 35 °F, the time required to effect good coagulation may increase considerably. For example, some studies indicate that about twice the time will be required at 34 °F as at 70 °F.

Table 8-1
Coagulation and Settling Tests

Sample Number	Cl_2 ppm	$FeSO_4$ gr/gal	$Ca(OH)_2$ gr/gal	Floc Formation	Settling Time (Min)	Super-natant
1	12	1	1	Poor	8	Hazy
2	12	1	2	Fair	7	Hazy
3	12	1	3	Fair	6	Hazy
4	12	2	1	Fair	7	Hazy
5	12	2	2	Good	5	Fair
6	12	2	3	Excellent	4	Clear

Coagulant Aids
There are times when the use of the conventional coagulants are not truly
conducive to good coagulation and settling. At such times, the use of chemicals
called polyelectrolytes and/or polymers can be used instead to enhance these
processes.   Under such circumstances, it is suggested that polyelectrolyte
manufacturers be contacted. They will conduct laboratory and field tests to
determine the proper applications.❑

CHAPTER 9

CONVENTIONAL TREATMENT PLANT

A conventional water treatment system, as distinguished from other demineralization treatments (e.g., reverse osmosis, electrodialysis, demineralization), is illustrated in Figure 9-1. It serves as the only treatment for many industries; and, at times, it is considered a necessary pretreatment for the demineralization procedures mentioned above. As noted in the figure, it consists of a reactor, chemical feeders, sand or coal filters, activated carbon purifier, and polishing filter. These systems operate at a constant flowrate, with the start-stop signal emanating from a treated-water storage facility.

Reactor
The principle on which the reactor is designed is that coagulation and flocculation are improved if they take place in the presence of previously precipitated particles (floc). To achieve this, it is necessary to build up a mass of previously precipitated solids, which are referred to as *slurry*. Note from Figure 9-2 that the raw water and chemicals enter the reactor in which coagulation, flocculation, and settling occur, each in its own compartment.

When the reactor is first placed into service, none of the settled solids are sent to waste. Rather, all of these solids are circulated back to the head end of the unit until the desired slurry is established. Depending upon the particular raw water and on the particular treatment, it may take from several days to several weeks to obtain a satisfactory slurry. Once this goal has been attained, the slurry concentration is maintained by the recirculation of only a portion of the settled solids, the remainder of which is discharged to waste. The settleability of the floc particles will depend on their source and on the type of treatment under consideration.

For example, if alkalinity and hardness reduction (lime treatment) is practiced, the resultant floc particles will consist mainly of calcium carbonate and magnesium hydroxide, of which the latter is itself a coagulant. Mixed with these solids are those that result from the use of a coagulant. These solids will be ferric or aluminum hydroxide.

This type of floc particle is perhaps the fastest settling that will be

73

encountered in water treatment. If, at the other end of the scale, the water is being treated only for the removal of color, colloidal particles, and/or organic matter, the resultant ferric or aluminum hydroxides are found to be the slowest settling particles encountered.

Table 9-1 will give some idea of settling velocities (at 59 °F) of the various floc particles likely to be encountered.

These data indicate that calcium carbonate floc settles from 3 to 5 times as fast as does the fragile floc. They are translated into a new parameter called the rise rate, which is measured in units of gallons per minute per square foot (gal/min/ft^2) of settling area. This settling occurs in the annular space of the reactor, whose vertical cross section is noted in Figure 9-3.

Operating experience has shown that the maximum rise rate should not exceed 0.75 gal/min/ft^2 when aluminum hydroxide or ferric hydroxide is the result of treatment. On the other hand, when practicing the lime or lime/soda softening that results in the heavier calcium carbonate and magnesium hydroxide floc, a rise rate of up to 1.5 gal/min/ft^2 is not unreasonable, although this is not practiced for product water use.

Also, it is possible to increase the rise rate through the use of plate or tube settlers, which are installed just below the settling area of the reactor. It is suggested that such an arrangement be discussed with the equipment manufacturer.

Following are the calculations required to design a reactor (such as the one shown in Figure 9-3) that will be capable of treating 300 gallons per minute (gpm):

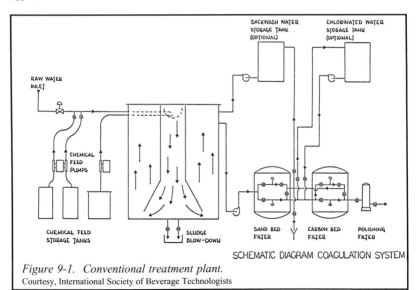

Figure 9-1. Conventional treatment plant.
Courtesy, International Society of Beverage Technologists

Settling area required = 300 gpm ÷ 0.75 gpm/ft² = 400 ft²

The annular area of the reactor tank has a width of 20 ft - 6 ft = 14 ft.
Length of the tank = 400 ft² ÷ (20 ft - 6 ft) = 28.57 ft

From the fact that the length and width of the reactor are now known, another parameter, can be determined. This is called *contact, detention* or *retention* time.

To assure the complete destruction of bacteria and other organic matter, it has long been the practice to allow a 2-hour contact time between chlorine and the water undergoing treatment. To calculate this parameter, it is necessary to determine the depth of liquid in the reactor. Using Figure 9-3 again as an example, the required volume of water in the reactor tank would be found as follows:

300 gpm × 2 hours × 60 min = 36,000 gal
36,000 gal ÷ 7.5 gal/ft³ = 4,800 ft³

Therefore, depth of the water in the reactor =

4,800 ft³ ÷ 28.57 ft long × 20 ft wide = 8.4 ft

Figures 9-2 and 9-3 do not show an explicit reactor design. Each such design

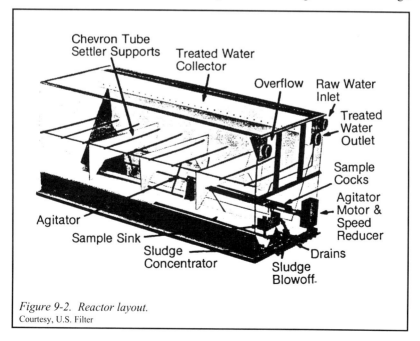

Figure 9-2. Reactor layout.
Courtesy, U.S. Filter

has its own operating and physical peculiarity, and each manufacturer will furnish necessary instructions for both. As a result, the tank itself could be from 9 to 10 ft deep.

Filtration
The water delivered to the sand or coal filter will contain from 5 to 10 ppm of suspended solids. These must be removed by the filter or they will seriously impair the efficiency of the activated carbon purifier, which is the next step in this process.

As shown in Figure 9-4, the filter is so constructed that the water being filtered passes through the filter bed from the top to the bottom, either by pressure or gravity. At the bottom of the filter shell, the filter water is collected by an underdrain and then passes on to the activated carbon purifier.

Different manufacturers suggest slightly different depths of filter media, but the following, reading from the bottom layer up, can be regarded as a general guide:

Gravel type	Diameter	Depth
Coarse gravel	3/4 - 1½ in.	8 in.
Medium gravel	½ - 3/4 in.	2½ in.
Fine gravel	¼ - ½ in.	2½ in.
X-fine gravel	1/8 - ¼ in.	3 in.
Torpedo sand	0.8 - 1.2 mm	3 in.
Sand	0.45 - 0.55 mm	20 - 24 in.

All material below the sand is called the support bed, which serves not only to keep the sand bed in place, but also aids in equally distributing the path of flow during filter and backwash stages.

The sand particles follow these general specifications:

Table 9-1
Floc Characteristics

Type of Floc	Application and Agent	Settling Velocity Feet/Minute
Fragile floc	Color removal - $Al(OH)_3$	0.12 - 0.14
Medium floc	Algae removal - $Al(OH)_3$	0.20 - 0.30
Strong floc	Turbidity removal - $Al(OH)_3$	0.24 - 0.35
Strong floc	Lime softening - $CaCO_3$	0.24 - 0.35
	Organic matter - $Fe(OH)_3$	
	Calcium carbonate solids	0.40 - 0.66

Courtesy, Infilco Degremont, Inc.

Effective size (0.45 millimeters [mm] to 0.55 mm) indicates that the size of the sand particles will fall between these units. Uniformity coefficient (u.c.), which is determined as follows:

$$u.c. = \frac{\text{theoretical size of sieve (mm) that will pass 60\%}}{\text{theoretical size of sieve (mm) that will pass 10\%}}$$

The u.c. value should fall between 1.65 and 2.00.

Graded anthracite coal can be used in lieu of sand and gravel, and its specifications are substantially the same as those for the sand and gravel.

The basic underdrain system, as shown in Figure 9-5, is made up of a header pipe and a series of laterals. The number of laterals is usually a third of the tank diameter in inches (e.g., a 48-inch unit would contain 48 ÷ 3 = 16 laterals, 8 attached to each side of the header pipe.

In all cases, small holes are drilled into the underside of each lateral. The total area of all these holes will be approximately 0.33% of the horizontal area of the filter bed. The size of headers and the size and number of laterals will all increase as the capacity of the filter increases. A different type of underdrain, the nozzle underdrain system, is becoming widely used. Nozzles, or strainers as they are

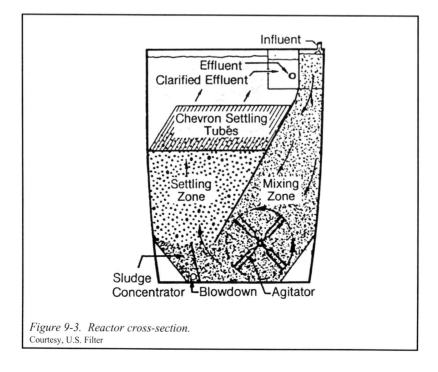

Figure 9-3. Reactor cross-section.
Courtesy, U.S. Filter

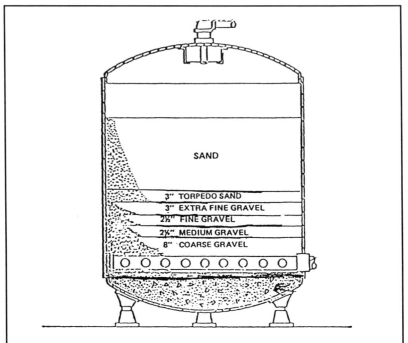

Figure 9-4. Pressure sand filter.
Courtesy, Infilco Degremont, Inc.

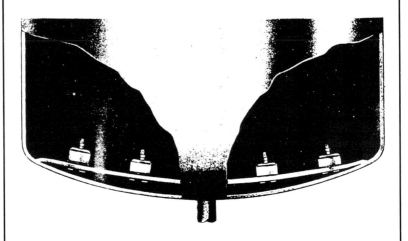

Figure 9-5. Underdrain.
Courtesy, U.S. Filter

sometimes referred to, are of various types and constructed of various materials. They can be installed in the laterals of the header and lateral system, or they can be installed in the flat bottom as shown in Figure 9-6. Pictured in Figures 9-7 and 9-8 are typical nozzles.

Nozzles are either slotted or wire-wound, with openings ranging from 0.007 to 0.025 inches. In such a design, a layer of sand deep enough to cover the nozzles is used. The openings in the nozzles will have to be small enough to contain the filter medium.

The capacity of a filter is commonly based on a flowrate of 2 gpm/ft^2, although rates of 3 or 4 gpm/ft^2 are not uncommon. This rate of flow determines the design flow for a 5-ft-diameter filter:

$$\text{Area} = \pi D^2/4 = \pi \times 5 \times 5/4 = 19.62 \text{ ft}^2$$

and at 2 gpm/ft^2:

$$19.62 \text{ ft}^2 \times 2 \text{ gpm/ft}^2 = 39.25 \text{ gpm}$$

Table 9-2 gives filter areas and their optimum flowrates. Figures 9-1 to 9-8 describe pressure filtration in which the water is forced through the filters by means of pumping, but the same parameters exist in gravity filtration (Figure 9-9). The gravity filter differs from the pressure filters in that only gravity is used to force the water through the filter bed. Gravity filters have the advantage in that the surfaces of the filter beds are visible, and that the backwash procedure

Table 9-2
Filter Areas and Flowrates

Diameter (in.)	Area (ft^2)	Filter Rate* (gpm)	Filter Rate* (ft^2)	Backwash Rate* (gpm)	Backwash Rate* (ft^2)
		2	3	10	15
36	7.06	15	23	75	115
42	9.6	20	30	100	150
48	12.56	25	38	125	190
54	15.9	32	48	160	240
60	19.6	40	60	200	300
66	23.7	48	72	240	360
72	28.3	56	84	280	420
78	33.2	66	99	330	495
84	38.5	77	117	375	555
90	44.2	88	132	440	660
96	50.3	100	150	500	750

*Rounded off

can be observed. On the other hand, gravity filtration requires an extra pumping to get the filtered water to its destination. Although Figure 9-9 illustrates a header-and-lateral underdrain system, nozzle underdrains and other types previously described are considered good practice.

Filter Operation

Filtration is the removal of any insoluble material through a porous medium.

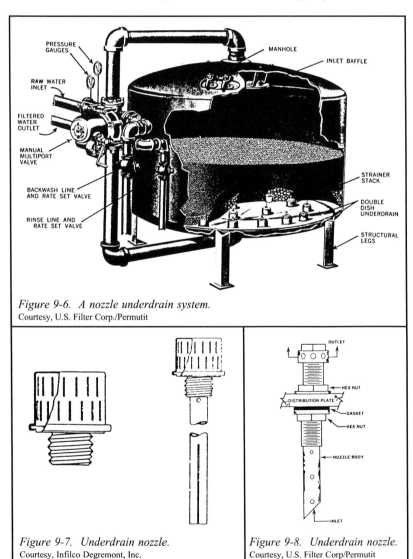

Figure 9-6. A nozzle underdrain system.
Courtesy, U.S. Filter Corp./Permutit

Figure 9-7. Underdrain nozzle.
Courtesy, Infilco Degremont, Inc.

Figure 9-8. Underdrain nozzle.
Courtesy, U.S. Filter Corp/Permutit

In the case of a conventional treatment plant, these are the solids not removed in the reactor.

The sand bed, which is supported by the underlying bed of graded gravel, is considered a porous medium. The size of the openings in this media is dependent on the effective size of the sand particles. The smaller the effective size, the smaller the openings and more efficient the filtration. That is to say, the smaller the sand particle, the smaller the openings in the media. However, a point is reached where too fine a sand will clog the filter too readily and, in consequence, give very short filter runs. It is on the basis of operating experience that the selected effective size is chosen.

When the filter is clean, the solid gel-like particles resulting from the coagulation and sedimentation will be retained by the top layer of sand particles. As soon as this layer of particles begin to clog, there is a head loss across the bed so that there is a greater pressure at the top of the bed than below. This difference in pressure will begin to push and penetrate the accumulated solids deeper into the bed, so that these filtered particles begin to occupy the spaces between the sand particles. If this process continues long enough, there will be a time when eventually some of the coagulated particles will completely penetrate the full depth of sand, and they will appear in the filter outlet. While this is occurring, the head loss across the filter bed will increase to the maximum allowable, so that little flow through the sand filter will be obtained. Before this time the filter will have to be backwashed in order to remove these particles

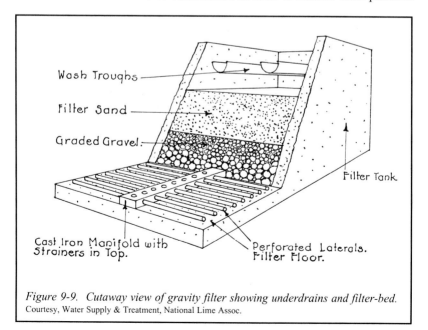

Figure 9-9. Cutaway view of gravity filter showing underdrains and filter-bed.
Courtesy, Water Supply & Treatment, National Lime Assoc.

from the sand bed and to start the filter cycle over again.

Sand filters are washed either daily or every other day. This practice, in a way, ignores the fact that a filter does not produce the best effluent until a coating of the coagulated particles *(schmutzdecke)* covers the surface of a sand bed. However, the daily routine of backwash is required in order to avoid the problem of solids penetration through the bed. In almost all municipal and industrial applications, filters are not washed on such a routine basis. Rather, the frequency of backwash is determined by the loss of head across the sand bed, usually 6 to 8 ft (2.5 to 3.5 lb). This loss of head at design flow is in addition to the loss of head through a clean filter bed.

To backwash a filter bed, water enters the underdrain system in the opposite direction from the flow of water during the filter cycle. The flow during backwash, then, is from the bottom of the filter bed upward through the sand bed and thence to waste. This can be directed back to the clarifier or to a backwash storage tank from which the backwash water is bled back into the clarifier at a reduced rate, or to drain.

The rate of backwash water, usually 6 to 8 times as great as the standard filter rate, must be sufficient to cause the sand bed to expand. While the sand bed is expanded to about 40% to 50% greater than its original volume, the individual sand particles are highly agitated. During this expansion and agitation, the particles will scour each other. This removes the gel-like floc from the sand particles. The floc is entrained in the flow of backwash water and removed from the filter.

The necessity of a properly designed underdrain system and of sufficient backwash rate can be seen by observing the action of the sand bed during backwash. First, however, the graded gravel serves the twofold task of supporting the sand bed, and also makes the distribution of the backwash water more even. While the backwash is flowing through the sand bed, it will be noticed that the sand appears to be boiling. That is, the sand will move upward in one portion of the tank and downward in another portion. Unless this action is minimized, some of the dirt and floc that has washed from the top layer of sand particles will attach itself to some of the sand particles at the bottom.

If this situation continues for any length of time, the buildup on those affected sand particles becomes large enough to cause what are called *mud balls*. If a sufficient number of these mud balls are allowed to form, the result will be *channeling*, which is the term for indicating that channels exist in the filter bed. In such a case, part of the sand bed has become ineffective as a filter media, and this will result in less efficient filtration since the same quantity of water is being filtered through a smaller volume of sand.

If it is believed that such a situation exists, extraordinary means of cleaning the sand beds must be used or the sand bed replaced.

To clean the sand bed and to get rid of mud balls, it will be necessary to agitate the sand using a rake or something similar, thus breaking up the mud balls, while the unit is being backwashed. This procedure will have to be performed several times in succession to be successful.

Good operating procedure is to replace the filter sand yearly, or at least every 2 years. Some facilities have records indicating that their gravel supporting beds have not been changed for 5 to 10 years. This is not good operating practice. Although it is true that there is no real attrition or wear and tear of the gravel particles, it is wise to remove and replace the gravel at least every 2 to 3 years in order to inspect the underdrain system and tank lining.

Activated Carbon Purifier
The design of the purifier copies in every detail the design of the filter. The one difference is that activated carbon (Figure 9-10) replaces the bed of sand.

The activated carbon effectively removes the excess chlorine that has passed

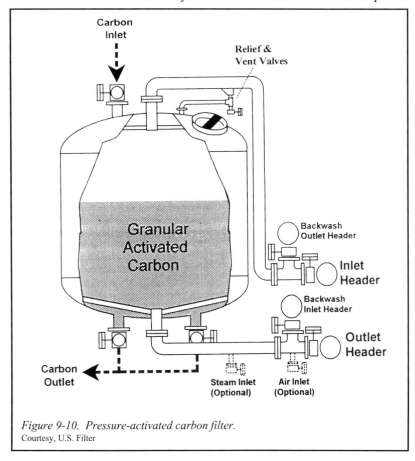

Figure 9-10. Pressure-activated carbon filter.
Courtesy, U.S. Filter

through the sand filter. The depth of the activated carbon will vary, depending not only on the flowrate, but also on the particular type and brand of carbon used. As an example, and assuming the standard filtration rate of 2 gal/min/ft², a 20-inch depth of one particular brand of granular activated carbon (GAC) was found to be sufficient for the removal of 8 ppm of chlorine residual for a period of 1 year (at five 8-hour days per week).

Other carbon manufacturers recommended similar but not identical quantities of GAC (check with the manufacturer). A 20-inch depth of GAC in a 4-ft-diameter shell will equate to the following volume:

$$\pi D^2/4 \times 20 \text{ in.}/12 \text{ in.} = 20.94 \text{ ft}^3$$

A flowrate in the same vessel at 2 gal/min/ft² of area =

$$\pi D^2/4 \times 2 = 25.1 \text{ gal/min (or gpm)}$$

These calculations are made in an effort to explain yet another parameter:

$$25.10 \text{ gal/min}/20.94 \text{ ft}^3 = 1.20 \text{ gal/min/ft}^3$$

Other carbons show some variations from the above. As a result, there seems to be a tacit approval to use a value of 1 gal/min/ft³ when designing the activated carbon purifier.

Under proper backwash rates, which may vary from 10 gal/min/ft² to 15 gal/min/ft², the GAC bed will expand. As with a pressure sand filter, the side sheet must be long enough to accommodate a 40% to 50% GAC bed expansion during each backwash.

Figure 9-10 indicates that this particular supportive bed depth totals 19 inches. The 20-inch-deep GAC would expand to about 30 inches during backwash so that the total bed depth would be 49 inches; therefore the length of the side sheet would probably be a minimum of 60 inches.

Carbon Tower

When greater quantities of product water are required, it becomes advantageous to use a modification of the unit described above. Since the efficiency of GAC is primarily dependent on the contact time of water and GAC, the vessel described above can assume the tower configuration shown in Figure 9-11.

Note in Figure 9-11 that the nozzle underdrain has eliminated the use of the supporting bed. If it is desired to rate the purifier at 4 gal/min/ft², then it is necessary to use only a 40-inch layer of carbon, which will maintain the same contact time. At 6 gal/min/ft² the GAC depth would be 60 inches.

A GAC tower would be used in order to save cost and space in the larger treatment plant.

If the activated carbon in both cases is the same material, the backwash rates remain the same. However, carbons produced by different manufacturers have different mesh sizes and different densities. It must be kept in mind that the smaller the mesh (the greater the mesh number), the lower the backwash water requirement. For some carbons, the backwash rate is the same as the backwash required for sand, 12 to 15 gal/min/ft^2. For the lighter carbons, a backwash rate of 12 gal/min/ft^2 would wash the carbon right out of the tank. Therefore, great care must be taken that the correct backwash rate be used for each particular activated carbon.

It should be pointed out that the greater the backwash rate, the greater the likelihood of sufficient scouring of the carbon particles. Purifiers should be backwashed daily to free the carbon particles of the compacting caused by the downward flow and also to redistribute the particles throughout the bed. Proper backwash flowrates in many instances will result in a substantial reduction in the bacteria count of the purifier effluent.

Carbon towers have a number of advantages:

● Since the horizontal cross-section areas of units of the same diameter are the same, one such tower can replace three of the older type (a 60-inch tower versus three 20-inch beds), resulting in a considerable reduction in space required.

● For a 60-inch tower, only one-third of the quantity of backwash water is required over that for three 20-inch beds, since the backwash rate remains the same.

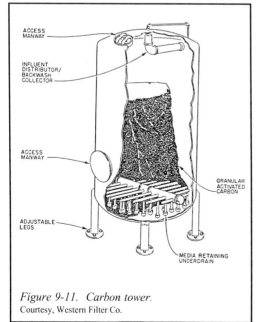

Figure 9-11. Carbon tower.
Courtesy, Western Filter Co.

● Stainless steel towers will each cost approximately $30,000 to $40,000 for the sizes likely to be required. But since only one-third the number of units of same diameter with the attendant equipment such as valves and controllers are required, the use of towers will result in a lesser cost.

As an example, assume that a flow of 1,000 gpm is required and further, using the 1-gal/min/ft^3 contact time. For flows of this magnitude, at least three units should be employed. Then:

$$1{,}000 \text{ gpm}/3 \text{ units} = 333.3 \text{ gpm/unit}$$

This would require 333.3 ft³ of GAC per unit. Now assume a 60-inch GAC bed depth (5 ft). Then the cross-sectional area of the vessel is determined as follows:

$$333.3 \text{ ft}^3/5 \text{ ft} = 66.7 \text{ ft}^2/4 \ = \pi D^2/4 \text{ and}$$
$$D^2 = 66.7 \times 4/\pi$$
$$D^2 = 84.9$$
$$D = 9.21 \text{ ft} = 9 \text{ ft, } 2\frac{1}{2} \text{ in.}$$

Expanded bed depth during backwash =
60 in. + 60 in. × 0.40 = 84 in. = 7 ft.
Side sheet should be approximately 8 ft.

Chemical Feed Systems
It is perhaps fortunate that practically all product water treatment plants are of such capacity that solution-type feeders can be used for feeding the required dosages. This type of feeder is reliable, easy to maintain, readily adjustable, and

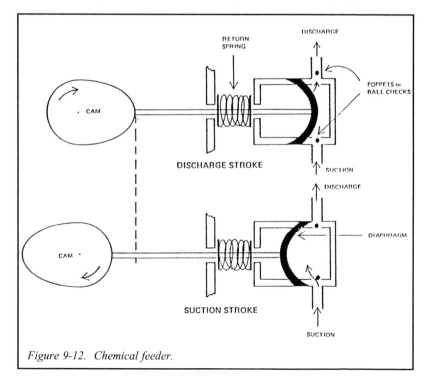

Figure 9-12. Chemical feeder.

can be constructed of chemical-resistant materials. The one exception to this statement is that the amount of chlorine required is such that in almost all cases, chlorine gas can be fed in the gaseous form. In those few cases in which chlorine gas is not applicable, the chlorine can be made into a solution. The principle on which these feeders are designed is illustrated in Figure 9-12. Note that a reciprocating motion is delivered to the diaphragm through the use of a cam.

During the discharge stroke, the shaft to which the diaphragm is attached rides against the high point of the cam. In so doing, the diaphragm is forced outward. As this occurs, pressure builds up in the chamber, causing the suction poppet or ball check to seat and the discharge poppet or ball check to open, thus causing a flow of chemical solution into the discharge line to the point of use. At the same time, the return spring is compressed.

As the cam rotates, the shaft to which the diaphragm is attached will move to the left, forced there by the return spring. As this occurs, the end of the shaft will eventually ride against the low point of the cam, and this would give zero flow. During this time, the discharge poppet or ball check would seat. At the same, the suction poppet valve would open to allow the chemical solution to fill the next discharge stroke.

The best-designed feeder system will consist of one feeder and one solution tank, as shown in Figure 9-13, for each chemical to be fed.

The solution tanks should be large enough to contain sufficient chemical to last for 24 hours or some other convenient time period. As an example, assume it is desired to feed 2 gr/gal of coagulant to a flow of 250 gpm.

Figure 9-13. Solution feeder assembly.
Courtesy, Infilco Degremont

Calculate the amount of coagulant required:

$$2 \text{ gr/gal} \times 250 \text{ gpm} \times 1,440 \text{ min/day}/7,000 \text{ gr/lb} = 102.8 \text{ lb/day}$$

Assume that the chemical feeder can pump up to 5 gal/h:

$$5 \text{ gal/h} \times 3,785 \text{ cm}^3/\text{gal}/60 \text{ min/h} = 315.42 \text{ cm}^3/\text{min}$$

Set feeder rate setter to 50%, so that any required adjustment can be up or down. Then:

$$315.42 \text{ cm}^3/\text{min} \times 0.50 = 157.7 \text{ cm}^3/\text{min}$$

Hence, the solution tank must have a capacity of at least the following:

$$157.7 \text{ cm}^3/\text{min} \times 1,440 \text{ min/day}/3,785 \text{ cm}^3/\text{gal} = 59.99 \text{ gal, rounded to } 60 \text{ gal}$$
$$60 \text{ gal} = 60 \text{ gal}/7.5 \text{ gal/ft}^3$$
$$= 8.00 \text{ ft}^3 = \text{content of tank}$$

Assume a tank diameter of 2 ft, 6 in. Then 2 ft has area = 3.14 ft^2 and liquid depth in this tank would be calculated as follows:

$$8 \text{ ft}^3/3.14 \text{ ft}^2 = 2.54 \text{ ft} = 30.5 \text{ in.}$$

This solution tank, as shown in Figure 9-14, should then be about 36 inches deep, which would be 5 in. above the approximately 31-in. liquid level.

Since 60 gallons of water will contain 102.8 lb of coagulant, each gallon would contain 102.8/60 = 1.71 lb/gal.

Now check Table 9-3, which indicates that ferrous sulfate, if it is the coagulant, is soluble to the extent of 2.2 lb/gal; hence, this concentration is acceptable.

Each inch of solution now contains 102.8 lb/31 in.

$$= 3.3 \text{ lb ferrous sulfate/in.}$$
$$= 3 \text{ lb, 4 oz/in.}$$

Thus, to recharge the solution tank it is necessary to measure the drop in level in the solution tank from the 31-in. mark; add 3 lb, 4 oz for each inch below the 31-in. mark; fill to this mark with water; and mix. Now, assume that it is required to increase the dosage from the original 2 gr/gal to 3 gr/gal. One way to do this would be to

Figure 9-14. Chemical solution tank.

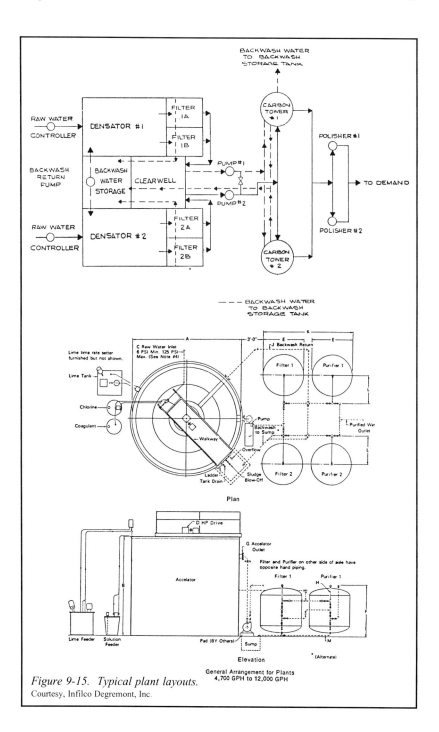

Figure 9-15. Typical plant layouts.
Courtesy, Infilco Degremont, Inc.

Table 9-3
Solubilities of Reacting Chemicals

Chemical	Solubility	Temperature
Aluminum sulfate	2.3 lb/gal	20 °C (68 °F)
	3.09 lb/gal	60 °C (140 °F)
Calcium hypochlorite	0.32 lb/gal	20 °C
Calcium chloride	6.0 lb/gal	20 °C
Ferric chloride	3.7 lb/gal	20 °C
Ferric sulfate	2.5 lb/gal	20 °C
Ferrous sulfate	2.2 lb/gal	20 °C
Lime (slurry - insoluble)	2.0 lb/gal	20 °C
Sodium aluminate	1.0 lb/gal	20 °C
Soda ash	3.0 lb/gal	20 °C

increase the pump setting from 50% (157.7 cm^3 per min) to 75% (236.55 cm^3 per min), but with this change, the chemical would last only 16 hours. Perhaps a better way to accomplish this is to recharge the tank as shown above and add an additional 50% more ferrous sulfate. (A raise to 3 gr/gal from 2 gr/gal is a 50% increase.)

Add 102.8 lb $\times$ 0.50 = 51.4 lb, and fill tank to 31-inch mark.

With an increase in the base charge in this manner rather than a change the original pump setting (50%), the dosage will have been as desired, and the tank contents will still last 24 hours.

It should be determined again that the increased base charge does not exceed the solubility of ferrous sulfate.

= (102.4 lb + 51.4 lb)/60 gal = 153.8 lb = 2.5 lb/gal

= 2.50 lb ferrous sulfate per gallon of water, which would be the absolute maximum concentration allowable.

Should the tank selection exceed the solubilities shown in Table 9-3, it becomes necessary to use a larger solution tank. To reduce the base charge, the operator would wait until the level drops in the solution tank, and proceed from this point.

For example, if it is desired to reduce the base charge from 2 gr/gal to 1 gr/gal, the operator could wait until the level in solution tank drops halfway (to 16½ inches), and refill with water to the 31-inch mark. Again the feeder setting need not be changed and the base charge would last the desired 24 hours.

See Figure 9-15 for layouts typical of conventional treatment plants.❏

CHAPTER 10

DISINFECTION AND STERILIZATION

A review of the previous chapters will attest to the fact that practically all of the substances discussed were of an inorganic nature (minerals). Note further that other minerals such as lime and coagulants were used for their removal. It is probably no overstatement to say, however, that the most or at least equally troublesome problems encountered in water treatment are of an organic nature (plant and animal life). As a result, a new glossary of terms is required for a discussion of the phenomena of disinfection and/or sterilization.

Disinfection - The use of a substance or process to destroy *harmful* bacteria and/or germs.

Sterilization - The use of a substance or process to destroy *all* bacteria and/or germs.

Microorganism - A very tiny living plant or animal that can be seen only through a microscope. Examples are bacteria (both harmful and harmless to human life), Protozoa, and viruses.

Bacteria - Very tiny, usually microscopic simple plants. Bacteria consist of single cells and under certain conditions multiply rapidly. Certain bacteria cause diseases such as fever, typhoid, and pneumonia. Other bacteria, such as the ones that turn cider or wine into vinegar, are useful.

Algae - A group of water-borne plant organisms. Some algae, such as the blue-green scum that is seen in pools of stagnant water, are single-celled. Other algae are extremely large and can be measured in inches or feet.

Germ - Microscopic plant or animal that causes disease.

Carcinogen - A substance that produces cancer.

Pathogen - A plant or animal that causes disease. A germ is a pathogen.

Nonpathogenic - Describes a plant or animal that does not cause disease.

Potable - Fit to drink.

Chlorination, superchlorination - The use of or addition of various amounts of chlorine to water.

Chlorine demand - That quantity of chlorine required to kill all living organisms (bacteria, algae) plus the quantity that will react with other substances such as iron or manganese, or other organic matter.

Residual chlorine - A measure of the chlorine that remains in the water after the chlorine demand has been satisfied.

Chloramines - Compounds that are a combination of chlorine and ammonia and that have a general formula, NH_xCl_y. Ammonia can be of organic origin or it can be as inorganic matter that is added to the municipal supply. As such, the stable chloramine formed assures a disinfecting ability in long transmissions such as in water mains.

Trihalomethane (THM) - A compound formed when a member of the halogen family (such as chlorine or bromine) reacts with a certain precursor present in the water.

Disinfection by-products (DBP) - These are products formed as a result of the disinfection and/or sterilization of water. As a rule, these are deemed to be cancer-causing in certain concentrations.

Except in emergencies such as water-main breaks or floods, it is safe to assume that the water received by the home or manufacturing plant has been disinfected. Disinfection will remove all harmful bacteria and germs (pathogens), and as such, it will deliver a potable supply for the general public and for some uses in the manufacturing plant.

Disinfection can be accomplished through the use of various chemicals such as chlorine, chloramine, chlorine dioxide, potassium permanganate, or ozone; or by a physical process such as ultraviolet light or the use of membrane filtration such as reverse osmosis.

Although each of these options can be used for disinfection, a survey reported in the *AWWA Journal* (American Water Works Association, Jan. 1988) indicates that chlorination is practiced by about 85% of municipalities. The remainder of municipalities use chloramine, chlorine dioxide, ozone, or potassium permanganate.

The purpose of such disinfection by the public supplier is to remove the harmful bacteria and viruses and thus to make the water potable. Industry, on the other hand, must further treat (sterilize) all bacteria, algae, and other organics that create a chlorine demand.

In other words, for industry to obtain the best product waters, the public supplies must be treated beyond sterility. To this end, industry must resort to a combination of the processes previously mentioned. Additionally, if chlorination is used to the fullest extent, it is necessary to practice *breakpoint chlorination,* as described in the following discussion.

Chlorination

Chlorination is the process in which chlorine gas (Cl_2) or a chlorine compound such as sodium hypochlorite (NaOCl) is added to the water for the purpose of disinfection and sterilization. In addition to the bactericidal action that chlorine imparts, these other benefits result from its use:

● Improved coagulation, resulting from the attack on the organic matter that may be present. It is difficult to get the best coagulation in the presence of organic matter.

● Reduction of taste-, odor-, and color-causing matter.

● Oxidation of iron and manganese that might cause color and deposits.

● Reduction and removal of any organic matter.

When chlorine is added to water, the following reaction occurs:

$$Cl_2 + 2H_2O \longrightarrow HOCl + H_3O^+ + Cl^-$$
$$HOCl + H_2O \longrightarrow H_3O^+ + OCl^-$$

These end products are strong oxidizing agents, and as such, will disinfect the water. However, it is stated in "Water Supply and Treatment" from the National Lime Association (1943) that a further reason than oxidation must be looked for alone to explain the disinfection. One plausible explanation is that chlorine unites, at least in part, with the cell structure of the organism to form chloro-products that act as toxic poisons to these organisms.

The amount of hypochlorous acid (HOCl) that is present in a reaction is dependent on the pH of the water under consideration. At a pH of 6.5 or less, almost all the chlorine is present in the form of HOCl; whereas at a pH of 9.0 or above, the ionized hypochlorite ion, OCl, accounts for all of the available chlorine. Since it is HOCl that is the principal disinfectant in the chlorine solution, the disinfective ability of chlorine is most effective at the lower pHs.

The residual chlorine required for simple disinfection is determined bacteriologically. Usually, a 0.2- to 0.5-ppm residual will suffice.

Breakpoint Chlorination

Figure 10-1 shows the effect of the addition of chlorine gas to distilled water.

(HOCl is a very unstable compound that dissociates as shown above.)

Water #1 is distilled water that contains no material that will react with chlorine. As a result, the free chlorine residual will increase proportionately as the amount of chlorine is increased. All of the residual will be as HOCl, which is available for disinfection. This is called *free available chlorine.*

Water #2, on the other hand, is of the type that is most likely to be encountered. It will probably contain iron, manganese, nitrogen (organic and/or inorganic), and other organic matter (humic, fulvic, and tannic acids), all of which will react with chlorine.

The following statement is excerpted from "Water Supply & Treatment" (National Lime Association, 1943).

"Recently a new technique in the control of chlorination has been instituted which affords a more successful use of chlorine for controlling tastes, odors, and bacteria. The term "break-point" chlorination is used in connection with this practice and is explained as follows:

On the addition of increasing doses of chlorine to a series of water samples the residual chlorine gradually increases to a point where further additions of chlorine are followed by a decrease in the residual. Finally on still further additions of chlorine the residuals again begin to increase. This second point is designated as the 'break-point'."

A more succinct definition of breakpoint chlorination is illustrated in Figure 10-2:

Phase I. When chlorine is added to water, there will be an "instantaneous" reaction with substances such as Fe^{2+} (ferrous) iron and Mn^{2+} (manganous) manganese, and perhaps also with some bacteria and algae.

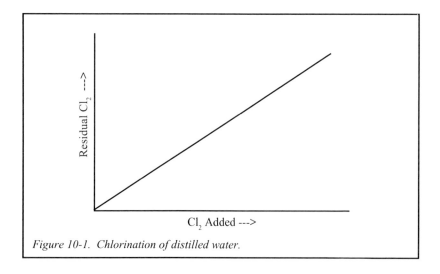

Figure 10-1. Chlorination of distilled water.

In this phase, iron and manganese will be oxidized and precipitated to the ferric (Fe^{3+}) and manganic (Mn^{3+}) state, and thus removed from further consideration.

Phase II. Further addition of chlorine will result in reactions with any of the nitrogenous compounds (ammonia) in the water. These reactions will result in the formation of chloramines (NH_xCl_y).

For example, when ammonia or an ammonia compound exists in the water:

$$NH_3 + HOCl \longrightarrow NH_2Cl + H_2O$$

Phase III. If more chlorine is added, it will tend to destroy most or all of the chloramines that were formed in Phase II. For example, when more chlorine is added:

$$NH_2Cl + HOCl \longrightarrow NHCl_2 + H_2O$$

Phase IV. Since all of the oxidizable substances (chloramines) have now been reacted with, any further addition of chlorine will result in the formation of free chlorine (HOCl), which is the desired disinfectant. This point is called the *breakpoint,* and the action here is similar to that shown previously when chlorine is added to distilled water.

In some cases it is impossible to destroy all of the chloramines, so that at any

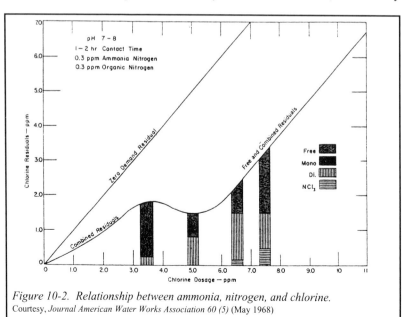

Figure 10-2. Relationship between ammonia, nitrogen, and chlorine.
Courtesy, *Journal American Water Works Association 60 (5)* (May 1968)

point beyond breakpoint there could be a combination of free chlorine and chlorine that is combined with a nitrogenous compound.

Note that chloramines were given the general formulation of NH_xCl_y. Table 10-1 shows in greater detail the different compounds of chloramine that are involved in the chlorination process.

Most natural (raw) waters contain both inorganic and organic nitrogen. Inorganic nitrogen occurs naturally and some is at times added for trihalomethane control by water suppliers such as municipalities, as will be discussed later. This nitrogen requires a short chlorine contact time (20 to 30 minutes).

Organic nitrogen occurs only naturally. Long chlorine contact time is required for removal. Reactions result in complex organic nitrogen compounds, probably never completely removed from the water.

As increasing amounts of Cl_2 are added to the water, a drop in the pH can be expected, and the following compounds will result:

At pH 8.5 or greater,
$$NH_3 + HOCl \longrightarrow NH_2Cl + H_2O$$
$$\text{Ratio } Cl_2:NH_x = \text{less than } 5:1$$

At pH 4.5 to 8.5,
$$NH_2Cl \text{ (monochloramine)} + HOCl \longrightarrow NHCl_2 \text{ (dichloramine)} + H_2O$$
$$\text{Ratio } Cl_2:NH_x = \text{greater than } 5:1.$$
$$\text{Dichloramine will impart a taste.}$$

At pH below 4.5:
$$NHCl_2 \text{ (dichloramine)} + HOCl \longrightarrow NCl_3 \text{ (nitrogen trichloride)} + H_2O$$
$$\text{Ratio } Cl_2:NH_x = \text{greater than } 15:1$$
$$\text{The } NCL_3 \text{ is an unstable gas with a foul, sharp, acrid odor.}$$

It is unlikely that NCl_3 will ever exist in actual practice, since it readily breaks down into its components.

Table 10-1
Chloramine Formation Equilibria

pH	% NH_2Cl	% $NHCl_2$	% NCl_3
4	5	25	70
5	65	35	0
6	80	20	0
7	95	5	0
8	100	0	0

Courtesy, *Ultrapure Water* journal (Sept. 1985)

In Figure 10-2, note from the curve and its legend that at breakpoint all chloramines except NCl_3 will exist, and that beyond breakpoint HOCl increases. Also note that both NH_2Cl and $NHCl_2$ exist beyond breakpoint. Also included beyond breakpoint could be other chlorinated organic compounds not destroyed at breakpoint.

The quantities of these compounds that remain in the treated water are called *residuals*.

Free available chlorine is that portion that exists as HOCl. This chlorine is available to react, and it has greatest bactericidal strength.

Combined available chlorine residual is that portion that has reacted with any of the ammonia or organic nitrogen compounds.

Total chlorine residual is the sum of free available chlorine and the combined chlorine residual.

Total chlorine residual = free chlorine + combined chlorine
Total chlorine residual = HOCl + chloramines

Although chloramines are long-lasting compounds, their disinfecting power is not as great as that of free chlorine. Further, it is difficult to remove all traces of them from the finished water. They can, in sufficient concentration, be detrimental to various products.

It becomes evident that a free chlorine residual will be attained only after sufficient chlorine has been added to reach the breakpoint.

This free chlorine residual will assure removal of bacteria, algae, color, taste, and odor. It is quite stable, and this water can be sent to process; or to other treatments, should this residual prove to be detrimental.

For example, in the carbonated beverage industry, a free residual of some 8.0 ppm is maintained for a period of 2 hours to assure complete destruction. The water is then dechlorinated through activated carbon beds, since any free chlorine would affect the taste and perhaps the color of the final beverage.

To determine the quantity of chlorine required in this instance, use the procedure described under "Jar Tests" in Chapter 8.

To the quantity of chlorine required to reach breakpoint, add an additional 8.0 ppm.

Assume this total is 20.0 ppm; then if chlorine gas is used (100% Cl_2) to treat 1,000 gallons per minute:

$$(1,000 \times 20) \div (17.1 \times 7,000) = 20,000 \div 119,700 = 0.167 \text{ lb } Cl_2/1,000 \text{ gal}$$

If NaOCl solution (12%) is used, each gallon contains $8.33 \times 0.12 = 1.0$ lb of Cl_2/gal. Hence, it would require 0.167 gal/1,000 gal of water. This is equivalent to 0.167 gal × 3,785 cm³/gal = 632 cm³/1,000 gal.

Practically all product waters demand the elimination of both free chlorine (HOCl) and chloramines (NH_xCl_y). These can be completely removed for all intents and purposes through the use of granulated activated carbon (GAC) and/ or powdered activated carbon (PAC).

These compounds can also be removed chemically as a result of their reactions with sulfur dioxide (SO_2):

$$Cl_2 + HOH \longrightarrow HOCl + HCl$$
$$SO_2 + HOH \longrightarrow H_2SO_3$$
and
$$HOCl + H_2SO_3 \longrightarrow H_2SO_4 + HCl$$
or
$$NH_2Cl\ (monochloramine) + H_2SO_3 + H_2O \longrightarrow NH_4HSO_4 + HCl$$

NH_4SO_4 is classified as an innocuous compound.

The dosage in the first case is approximately 1.0 ppm of SO_2 per 1.0 ppm of HOCl. The dosage for the removal of NH_2Cl will be approximately 2.0 ppm to 5.0 ppm SO_2 for most waters under consideration.

Sulfur dioxide will be fed in a method similar to that of feeding chlorine gas.

Precursors

Chlorine will also react with yet another group of organics that are usually found in surface supplies and, at times, in underground sources. Chief among these are humic, fulvic, and tannic acids, each of which is composed of hydrogen, oxygen, and carbon, and each of which is characterized by molecular weights in the thousands. They all will impart color, taste, and odor. When these compounds react with chlorine, the result is a compound called *trihalomethane* (THM). As a consequence, they are called *precursors* (forerunner or predecessor). Further, for ease of discussion they will be given the general formulation of methane (CH_4).

It would greatly simplify the treatment of water containing such substances if they were removed before chlorination, but it would prove to be both costly and difficult to do so. The most efficient means of removal of precursors at present are reverse osmosis, nanofiltration, and activated carbon.

At best, coagulation plus sand filtration is not fully effective. Ion-exchange processes, ultraviolet, aeration, or other disinfectants will remove little if any of these precursors.

Trihalomethanes

Trihalomethanes are by-products resulting from the chlorination of water supplies that contain organics. The general equation for the reaction is given by:

$$CH_4 + 3Cl_2 \text{ ----> } CHCl_3 + 3HCl$$
methane disinfectant   chloroform

The term *trihalomethane* is derived from the fact that three ("tri") atoms of hydrogen in the above reaction are replaced by three atoms of chlorine. The chlorine is one of a group of chemicals known as halogens, which also includes bromine and iodine among others; hence "halo." Finally, the molecular structure of the final compound is that of methane.

Other compounds that might be encountered in this chemistry are bromoform ($CHBr_3$), iodoform (CHI_3), and dichloroiodomethane (CH_2Cl_2I). The combinations of atoms can make a long list of such compounds, but suffice it to say that our interest in such compounds lies in the fact that they have been deemed by the authorities to be cancer-causing (carcinogens) when they exceed 100 parts per *billion*. It therefore becomes necessary for municipalities to exercise some control to limit the amount it sends to its customers. Chloramines and THMs are also classified as disinfection by-products (DBP). Note also from what follows that each of the disinfectants produces by-products of its own. These, including THM, are deemed to be carcinogenic beyond certain concentrations.

Chlorine Dioxide

Chlorine dioxide (ClO_2) exists as a gas, and it does not hydrolyze as does chlorine. Note the following reactions that take place in an alkaline solution:

$$ClO_2 + 2OH^- \longrightarrow ClO_3^{2-} + H_2O$$

The chlorite from this initial reaction is an oxidant, and a further reaction takes place:

$$ClO_2 + \text{organics} \longrightarrow ClO_2^- + \text{oxidized material}^+$$

Chlorine dioxide gas cannot be compressed, and hence must be manufactured on-site:

a) Chlorine, either as a gas or liquid, is mixed with sodium chlorite.

$$2NaClO_2 + Cl_2 \longrightarrow 2ClO_2 + 2NaCl$$

b) Sodium chlorite will react with hydrochloric acid.

$$NaClO_3 + 2HCl \longrightarrow ClO_2 + NaCl + H_2O$$

Every effort is made to assure the absence or minimum of free chlorine in the final product. Any free chlorine will result in unwanted chloramine and/or THM in the water under treatment.

Chlorine dioxide is a very stable disinfectant, and it is effective in the coagulation process. It does not form THM, chloramine, or chlorophenol. Its use is excellent for taste and odor control. Chlorine dioxide is also an excellent oxidant that can be used for the removal of iron and manganese.

The shortcomings of ClO_2 use are its need for on-site production and the fact that it does not remove all organics. Further, it produces chlorites and chlorates (DBPs), which are or may be suspect as carcinogens. It is difficult to state the required ClO_2 dosage. This is best determined by jar tests, and by bacteriological and residual tests (as in the use of chlorine).

Ozone

Ozone (O_3) is probably the strongest oxidizing agent available for water treatment. Although it is widely used throughout the world, it has not found much application in the United States. Ozone is obtained by passing a flow of air or oxygen between two electrodes that are subjected to an alternating current in the order of 10,000 to 20,000 volts.

$$3O_2 + \text{electrical discharge} \longrightarrow 2O_3$$

Liquid ozone is very unstable and can readily explode. As a result, it is not shipped and must be manufactured on-site. Ozone is a light blue gas at room temperature. It has a self-policing pungent odor similar to that sometimes noticed during and after heavy electrical storms. In use, ozone breaks down into oxygen and nascent oxygen.

$$O_3 \longrightarrow O_2 + O$$

It is the nascent oxygen that produces the high oxidation and disinfection, and even sterilization. Each water has its own ozone demand, in the order of 0.5 ppm to 5.0 ppm. Contact time, temperature, and pH of the water are factors to be determined.

Ozone acts as a complete disinfectant. It is an excellent aid to the flocculation and coagulation process, and it will remove practically all color, taste, odor, iron, and manganese. It does not form chloramines or THMs, and while it may destroy some THMs, it may produce others when followed by chlorination.

Ozone is not practical for complete removal of chlorine or chloramines, or of THM and other organics. Further, because of the possibility of formation of other carcinogens (such as aldehydes or phthalates) it falls into the same category as other disinfectants in that it can produce DBPs.

Potassium Permanganate

Although potassium permanganate ($KMnO_4$) is a more powerful oxidant than chlorine, it is purportedly less effective as a disinfectant. Its widest application in water treatment is as a regenerant of greensand exchange beds for the removal of iron and manganese.

$$3Fe(HCO_3)_2 + KMnO_4 + 7H_2O \longrightarrow MnO_2 + 3Fe(OH)_3 + 2KHCO_3 + 5H_2CO_3$$
$$3Mn(HCO_3)_2 + 2KMnO_4 + 2H_2O \longrightarrow 5MnO_2 + 2KHCO_3 + 4H_2CO_3$$

Recent interest in the use of potassium permanganate may be accounted for by the fact that it may result in the destruction of a small amount of precursors, but will not form THMs. However, its use as a disinfectant in water treatment remains at a minimum in this country.

Normal dosages vary from water to water and will be determined by bacteriological testing.

Ultraviolet Radiation

Figure 10-3 depicts the entire gamut of electromagnetic radiation as it is presently known.

The enormous temperatures on the sun create ultraviolet (UV) rays in great amounts, and this radiation is so powerful that all life on earth would be destroyed if these rays were not scattered by the atmosphere and filtered out by the layers of ozone gas that float some 20 miles above the earth.

This radiation can be artificially produced by sending strong electric currents through various substances. A sun lamp, for example, sends out UV rays that when properly controlled result in a suntan. Of course, too much will cause sunburn.

The UV lamp that can be used for the disinfection of water depends upon the

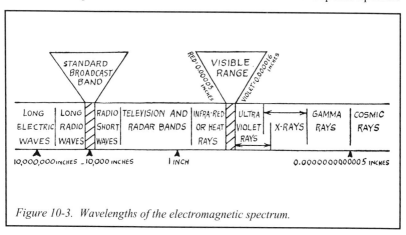

Figure 10-3. Wavelengths of the electromagnetic spectrum.

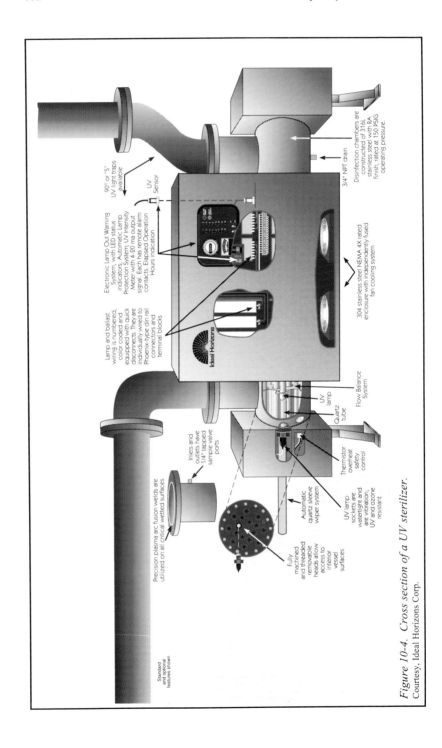

Figure 10-4. Cross section of a UV sterilizer.
Courtesy, Ideal Horizons Corp.

use of a low-pressure mercury vapor lamp to produce the ultraviolet energy. A mercury vapor lamp is one in which an electric arc is passed through an inert gas. This in turn will vaporize the mercury contained in the lamp; and it is a result of this vaporization that UV rays are produced.

The lamp itself does not come into contact with the water. The lamp is placed inside a quartz tube, and the water is in contact with the outside of the quartz tube. Quartz is used in this case since practically none of the UV rays are absorbed by the quartz, allowing all of the rays to reach the water. Ordinary glass cannot be used since it will absorb the UV rays, leaving little for disinfection.

Figure 10-4 indicates the water flow around the quartz tube. The UV sterilizer will consist of a various number of lamps and tubes, depending upon the quantity of water to be treated.

As water enters the sterilizer, it is given a tangential flow pattern so that the water spins over and around the quartz sleeves. In this way the microorganisms spend maximum time and contact with the outside of the quartz tube and the source of the UV rays.

Table 10-2
Removal of Disinfection By-products

Disinfectant	Disinfectant By-product	Disinfectant By-product Removal
Chlorine (HOCl)	Trihalomethane (THM)	GAC, resins, controlled coagulation, aeration,
	Chloramine	GAC-UV
	Chlorophenol	GAC
Chloramine (NH_xCl_y)	Probably no THM Others?	GAC UV?
Chlorine dioxide (ClO_2)	Chlorites Chlorates	Use of Fe^{2+} in coagulation, RO, ion-exchange
Permanganate ($KMnO_4$)	No THMs	
Ozone (O_3)	Aldehydes, carboxylics, phthalates	GAC
Ultraviolet (UV)	None known	GAC

The basic design flow of water of certain UV units is in the order of 2.0 gpm for each inch of the lamp. Further, the units are designed so that the contact or retention time of the water in the unit is not less than 15 seconds. Most manufacturers claim that the UV lamps have a life of about 7,500 hours, which is about 1 year's time. The lamp must be replaced when it loses about 40% to 50% of its UV output; in any installation this is determined by means of a photoelectric cell and a meter that shows the output of the lamp. Each lamp is outfitted with its own photoelectric cell, and with its own alarm that will be activated when the penetration drops to a preset level.

Ultraviolet radiation is an excellent disinfectant that is highly effective against viruses, molds, and yeast; and it is safe to use. It adds no chemicals to the water, it leaves no residual, and it does not form THMs. It is used to remove traces of ozone and chloramine from the finished water. Alone, UV radiation will not remove precursors, but in combination with ozone, it is said to be effective in the removal of THM precursors and THMs.

The germicidal effect of UV is thought to be associated with its absorption by various organic components essential to the cell's functioning.

For effective use of ultraviolet, the water to be disinfected must be clean, and free of any suspended solids. The water must also be colorless and must be free of any colloids, iron, manganese, taste, and odor. These are conditions that must be met. Also, although a water may appear to be clear, such substances as excesses of chlorides, bicarbonates, and sulfates affect absorption of the ultraviolet ray.

These parameters will probably require at least filtration of one type or another. The UV manufacturer will of course stipulate which pretreatment may be necessary.

Summary
Table 10-2 summarizes the foregoing section, and indicates that most of the disinfectants will leave a by-product that is or would possibly be inimical to health. This may aid with a decision as to whether or not precursors should be removed before these disinfectants are added to water.

If it is decided that removal of precursors is needed, research to date indicates that this removal can be attained through the application of controlled chlorination plus coagulation and filtration, aeration, reverse osmosis, nanofiltration, GAC or combinations of other processes.

In conclusion, the literature is replete with the pursuit of newer processes, or combination of known processes, or the search for new methods for removal of the DBPs. These will all require laboratory, pilot plant, and field experience before they can be fruitful, but as regulatory restrictions draw tighter, this research will be more than a "labor of love."❑

ORGANIC MATTER

Matter is "what things are made of," and all matter is categorized as either mineral, animal, or plant. Since there is a differentiation among these three categories, a further classification is possible. The *mineral* substances are called *inorganics*, while the remaining two, animal and plant substances, are termed *organic*.

The term *organic* arose from the fact that the first carbon compounds studied were all obtained from the bodies of plants or animals, that is to say, from *organized* or living things. For the most part, all plant and animal bodies are composed of carbon compounds, and all the processes by which they are formed during the growth and decay of living things are reactions of organic chemistry.

Today an organic substance is broadly defined as one that contains carbon. However, carbonates (CO_3), carbon dioxide (CO_2), and many other such compounds that contain the element carbon really arise from the mineral world, and consequently are relegated to the field of inorganic chemistry.

The rapid advance of chemical knowledge and the effort that has been put into the synthesis of new compounds have resulted in an almost endless number of organic compounds, so that it would probably be safe to say that there are over 1 million chemical compounds and substances that could be categorized as *organic*. Probably one of the most pervasive types of substance of concern to water quality control is organic matter. Consequently, it becomes necessary to make an attempt to determine which of the organics, or at least which class or group, are troublemakers.

Some considerations have been truly complicated by what we are now beginning to realize was a fairly straightforward problem. For example, the knowledge that the chlorination of a surface supply can result in the formation of a trihalomethane (THM), which could be chloroform and which is carcinogenic, has brought a new dimension to quality control.

As a result it becomes necessary, in order to discuss organic matter, to have two subclassifications, namely *nonvolatile organic substances* (NVOS) (Table 11-1) and *volatile organic compounds*(VOC) (Table 11-2). Examples of the former are algae and/or bacteria, about which we have been the most knowledgeable. Also, this classification includes other organics that have entered some water supplies as a result of pollution.

In order to have an understanding of the effect that each of certain substances can have upon the quality of a product, the following sections, sometimes brief, will attempt to describe them. The study of such substances is in large part the highly specialized field of microbiology, and the author can only introduce the reader to this information.

Rain and snow fall to the earth to furnish our eventual water supply. While in contact with the soil, this water picks up or absorbs various minerals, and also plant and animal life and waste products that have become a part of our environment. This organic matter becomes a part of the water supply, and it remains for municipalities to remove any substances that are harmful to human health.

Organic matter can exist as dissolved material, or as microscopic particles. It can be *pathogenic* (harmful to health) or *nonpathogenic* (not harmful to health).

It should be kept in mind that in making water potable, a municipality is not required to remove all of these organic substances. It does not do so simply because making a "sterile" water is too expensive. Further, even if a sterile water were obtained, it would undoubtedly become recontaminated to a degree in the transmission system.

The organic matter that always remains after such treatment consists basically of those bacteria and/or algae that are harmless to human life. However, these same substances are the source of many problems.

Table 11-1
Nonvolatile Organic Compounds

Substance	Treatment
Algae	Chlorine and other oxidizing agents plus coagulation, settling and filtration. Microstraining.
Plankton	Same as algae.
Detergents	Heavy aeration or ozone and activated carbon.
Phenols and phenolic compounds	Chlorine dioxide, ozone plus activated carbon.
Pesticides	Ozone and activated carbon.

Courtesy, U.S. EPA

Although most of the substances that will now be defined are not present in our drinking water, it is paramount that there be knowledge of the possibility that they might be a source of concern.

Nonvolatile Organic Substances (NVOS)
These do not readily change into a vapor, and do not evaporate.

Algae belong to the plankton family. Plankton are the tiny animal and plant organisms that are usually seen floating or drifting on the surface of water. These organisms are a form of plant life, distinguished by the fact that they contain chlorophyll, which imparts the green color to the plants. In size, they range from microscopic to hundreds of feet long. Of course, those algae in which manufacturers are interested range in size up to perhaps 20 microns (μm) (1/1,270 inches).

Algae are simple forms of plant or plant life organisms found in lakes,

Table 11-2
Volatile Organic Compounds

Halogenated Purgeables

Chloromethane	Bromomethane
Dichlorodifluoromethane	Vinyl chloride
Chloroethane	Methylene chloride
Trichlorofluoromethane	1,1-Dichloroethene
1,1-Dichloroethane	trans-1,2-Dichloroethene
Chloroform	1,2-Dichloroethane
1,1,1-Trichloroethane	Carbon tetrachloride
Bromodichloromethane	1,2-Dichloropropane
trans-1,3-Dichloropropene	Trichloroethene
Dibromochloromethane	1,1,2-Trichloroethane
cis-1,3-Dichloropropene	2-Chloroethyl vinyl ether
Bromoform	1,1,2,2-Tetrachloroethane
Tetrachloroethene	Chlorobenzene

Aromatic Purgeables

Benzene	Toluene
Ethylbenzene	1,2-Dichlorobenzene
1,3-Dichlorobenzene	1,4-Dichlorobenzene
o-Xylene	m-Xylene
p-Xylene	

Courtesy, U.S. EPA

reservoirs, or ponds. These organisms depend chiefly on sunlight for their existence, are mostly microscopic in size, and there is great variation in their physical structure. They are seen as the bluish-green or brownish slimes and froths at the edges of quiescent bodies of water.

In surface water, these microorganisms are rarely present in any amount in the cold water of winter. As the water warms up in the spring and summer, they start to grow and multiply at an unbelievably high rate. These particular species are not, as such, of great significance from a health standpoint, but they are the troublemakers that are met in the treatment and preparation of water for potable use. It has become common practice to "neutralize" them by disinfection through such means as chlorine as a first step in the coagulation phase of water treatment.

In addition to plant and animal organisms being present as solid particles, these can also exist in water as dissolved coloring matter—a true solution. They may also be present in a finely suspended state as a colloidal form. Dissolved organic matter is the end product of the natural life process of microorganisms, and usually shows up as oils or other soluble materials.

There are now several tests that will give definite indication whether or not a certain water contains enough of a certain organism to cause problems, but only actual experience in the production plant will resolve this question. The prediction as to whether or not objectionable organic matter is present is difficult because of the varying compositions of the water supply. It has been experienced, for example, that surface supplies are generally at their worst during the spring and fall turnovers in lakes and reservoirs.

A *turnover* is a seasonal phenomenon that occurs because of the temperature change that takes place in a body of still water. During the fall, the water in the upper portion of these waters cools. As it cools, its density increases, and this mass of colder water will settle to the bottom. The resulting thermal currents cause the warmer portion of the water at the bottom to rise. This water has been in contact with the vegetation that lies on the bottom, and hence contains much organic matter, which at this point will have been distributed throughout the lake and reservoir.

A similar turnover takes place during the springtime. As the ice that covers the surface of the water melts, the temperature of the water on the surface will rise from 32 °F to just below 40 °F, where it reaches its greatest density. This heavier water will tend to sink, causing the same action as described above.

It is a fact of life that these algae can exist in great numbers. Further, experience show that much can remain in the municipal water after its treatment. To remove such algae, the following processes are among those presently being used:

● The most common system incorporates disinfection, coagulation, sedimentation (with a 2-hour retention time), followed by granular filtration (sand or coal). This system removes not only algae but also does an excellent job on alkalinity reduction (if required) and bacteria removal.

● Membrane filtration (ultrafilter or reverse osmosis) is used; but the danger of fouling or plugging of the membrane is great if membrane filtration is used alone. Some type of pretreatment (see Chapter 13, "Filtration") is required.

● Diatomaceous earth is used, with the same caveats as above.

● In some instances, chlorination followed by filtration may deliver the desired effluent if there is not too much algae to be removed.

Bacteria are single-cell microorganisms, so classed because they can be seen only through a microscope. They range in size up to about 1.0 μm (1/1,000,000 of a meter). Bacteria differ in one respect from algae in that they contain no chlorophyll and hence no coloring matter.

Bacteria multiply by simple division, and if their carrier is not properly sanitized, the growth can reach a point reported as *too numerous to count* (TNTC).

The same treatments described under algae removal will deliver an almost 100% destruction of bacteria. In addition, ultraviolet is being used more frequently as a final step to assure elimination of bacteria.

Yeasts are also single-cell microorganisms, except that they are usually much larger in size than bacteria. They are of a yellowish color and, as a mass, form a "frothy" substance.

Yeasts multiply in the presence of starch and sugar to eventually form alcohol. Before the advent of fruit-juice-based beverages several years ago, yeast problems had been almost obliterated in beverage plants. Since yeasts are found in fruits, it is likely that they enter the plant in this manner. Yeast contamination has resurfaced, and the only palliative is strict sanitation of all equipment. It is presently surmised that failure to do so will continue to account for the greatest number of spoilages.

Mold consists of a furry growth on the surface of other organic solids. It is basically a fungus growth that takes place on damp or decaying substances. A mold contamination requires strict sanitation.

Sources of NVOS contamination. In this discussion of algae, bacteria, yeasts, and molds, the inference may be that this organic contamination stems only from the water source. However, experience shows that this is not always the case. Especially when there is a yeast contamination, there are other sources. Plant personnel can attest to the fact that likely sources can be found in places such as bottle-washing equipment and syrup and food-handling equipment. One likely source is in any dead leg in any equipment, pumps, or piping in the system. It is important to look at all of these possibilities and to maintain a high degree of sanitation throughout the entire plant.

Less-Common NVOS

While bacteria and algae will probably be present to one degree or another in the water supplied to a manufacturing plant, the following NVOS can be but are not likely to be present in these same supplies. The following are most apt to be present when the manufacturing plant uses its own wells.

Phenols and phenolic compounds. Phenols, per se, are a hallmark of pollution and are carcinogens. It is unlikely that any taste or odor would result if the phenol content is below 1 to 2 ppm. However, should any of the chlorophenol combine with chlorine to form chlorophenols, a taste and odor problem would result when the chlorophenols are present in amounts of 0.01 ppm (10 ppb). Remove through the use of GAC.

Hydrocarbons are the result of oil products (such as gasoline, kerosene, or fuel oil), and are the result of pollution. This pollution can take place in both surface and well supplies; and taste, odor, and toxicity can be the result.

The threshold for taste, odor, and carcinogenic harm from hydrocarbons is low, on the order of 200 to 300 ppb. These can best be removed using GAC.

Detergents. The introduction of biodegradable detergents has mostly eliminated any future pollution by conventional detergents, but remnants of the old nonbiodegradables still exist. These left the soapy taste that at one time was not too unfamiliar in water.

Pesticides are compounds used to control those organisms that are harmful to the plant and animal life needed to sustain humanity. Unfortunately, they in themselves are harmful to life and are pollutants. Pesticides, especially the chlorinated types, are a poison and are toxic to all forms of life. Perhaps less importantly, they do cause taste and odor in water when present in threshold amounts of less than 100 ppb.

Giardia **cysts** are microorganisms that are also classified as parasites. A cyst is an abnormal growth, containing diseased matter, in plants or animals. A parasite is an animal or plant that lives on or with another from which it gets its food. When *Giardia* cysts are present in drinking water, they come from fecal matter, and cause nausea, cramps, chills, and intestinal disorders very similar to severe diarrhea. They are strongly resistant to chlorine or chloramines except under very controlled conditions, and because of their small size (± 10 μm), good coagulation and filtration are required for their removal.

This is discussed further in Chapter 13, "Filtration," under "Diatomaceous Earth."

Volatile Organic Compounds (VOC)

A volatile substance can readily change to a vapor and will evaporate at ordinary temperatures. In contrast to the NVOS, volatile organic substances are soluble compounds that are loosely held in water, and as a result, can be vaporized or evaporated fairly easily (see "Air-Stripping").

It is important to note that VOC can be found in both surface and well supplies. Though legislation requires that these substances be removed from the water supplied to the public, they should not (but for one exception, THM), be of any concern to most manufacturing plants. However, since this utopia is not yet with us, it would be well for quality control to be aware of the facts as we know them today. The chief sources of these compounds are in those public supplies that are not treating "tainted well water," or from contaminated wells that are owned by the manufacturing plant.

Attempting to stay abreast of federal regulations for drinking water requires a full-time effort not only of water purveyors, both large and small, but also eventually by those corporate interests whose products use the water supplied to them. These regulations all result from the Safe Water Drinking Act (SWDA) of 1974, with amendments proposed in 1986. It is the intent of the EPA to cover these regulations in phases.

Phase I includes regulation applying to eight volatile organic compounds (VOC) that are listed in Table 11-3. Phase II includes regulations applying to chemical and microbiological compounds and pesticides. Some inorganic chemicals such as lead and mercury are included.

Microbiological contaminants include coliforms, parasites, viruses, *giardia*

Table 11-3
Maximum Level of VOC - U.S. EPA

Compound	ppb
Trichloroethylene	5
Carbon tetrachloride	5
Vinyl chloride	2
1,2-Dichloroethane	5
Benzene	5
para-Dichlorobenzene	75
1,1-Dichloroethylene	7
1,1,1-Trichloroethane	200

cysts, cryptosporidium, and turbidity. These five types of microbiological contaminants are all concerned with bacteria and other organisms that may cause infection and disease from contaminated and inadequately treated drinking water.

The EPA has stated that the law requires that recommended levels be set at a point that would result in no known or anticipated adverse health effects, with an adequate margin of safety. Under these proposals, the levels allowed for "probable" carcinogens are set very low. The EPA further states that the proposed goals are a significant step in the process of protecting the nation's public health from drinking-water contamination.

To give one special example, the key elements of the disinfectant by-product (DBP) regulators are as follows:

1 An MCL (maximum contaminant level) will be set for disinfectants and their residuals.

2 For chlorination by-products, including the reduction of THMs, MCLs will be in the range of 25 to 50 ppb.

3 There will be MCLs for ozonation by-products.

4 The EPA has tentatively determined that the most cost-effective approach to the D/DPB rule will be to set MCLs for a few contaminants that will serve as surrogate alarms for the overall safety of the water.

Such goals place a great challenge and responsibility on purveyors who supply large urban areas. The reader is reminded of the statement made in Chapter 1, "Water Sources," as to how zealously and expertly areas such as New York City guard the runoff water that fills their reservoirs.

Imagine, if you will, the cost of filtration (and maybe activated carbon treatment) of approximately a billion gallons per day if such were required.

Interestingly enough, some members of the U.S. Congress who originally participated in the enactment of pertinent legislation have been advised by their constituencies that some of the legislation is either too costly or too difficult to carry out, or both. As a result, some members of Congress are looking again at some parts of the legislation. There is the possibility of some changes in some of the existing requirements.

Under law, periodic testing of municipal suppliers must be strictly adhered to. As a result, this supply will be constantly monitored.

The latest U.S. EPA standard, in a regulation installed in 1988, now stipulates that the following VOC components cannot exceed certain amounts, as shown in Table 11-3.

If air-stripping does not deliver the above parameters, its effluent, as a rule, can be passed through GAC beds for almost 100% removal.❏

CHAPTER 12

GRANULAR ACTIVATED CARBON

Activated carbons are manufactured from a number of materials such as coconut, woods, sawdust, lignite coal (which resembles peat), and even from industrial by-products such as a residue from paper-pulp mills. Whatever the source of raw material, the process of manufacture consists of subjecting the raw material to steam temperatures of from 600 °C to 800 °C, usually in a rotary kiln under pressure. The products of combustion such as volatiles and some hydrocarbons are driven off, and it is this action that gives the carbon particle its distinctive structure.

Under a microscope an activated carbon particle is seen as having a structure quite similar to a sponge. This structure is made up of many pores, giving it great porosity. This structure is so composed that the walls forming the openings result in a tremendous surface area (see Figure 12-1). It has been estimated that 1 lb of activated carbon has a surface area of some 40 to 50 acres and will react with 3 to 5 lbs of free chlorine as Cl_2.

After formation, the carbon particles are pulverized and ground to various specified meshes, acid-washed, rinsed, air-dried, and graded to specification. *Mesh* is defined as "any of the open spaces of a screen or sieve." That is, a 50-mesh sieve is one with 50 open spaces per inch, and the term *mesh* is sometimes used alternatively with the term *particle size.* The mesh of activated carbon most widely used will range from 20 × 50 to 12 × 40.

A description of carbon as having a mesh of 20 × 50 indicates that practically all of the carbon particles will pass through a 20-mesh sieve, and practically all of this same carbon will be retained on a 50-mesh sieve. The percentages passed and retained on the various sieves will vary from carbon to carbon, and for specific figures, the manufacturer should be consulted.

It is noted that the larger the mesh, the finer the carbon. That is to say, a carbon of 20 x 50 mesh will be finer than a carbon of 12 × 20 mesh.

The ability of carbon to remove taste, odor, color, and undesirable organic matter from potable water is well known. Carbon was originally used and continues to be used in many municipal water treatment plants for such purposes. In this particular context, it is applied in powdered form; and it is fed either into the coagulation-sedimentation reactor, or into the reactor effluent that flows to the filters.

In contrast to this practice, the water supply to a manufacturing plant is heavily chlorinated to remove these undesirable components, and granular activated carbon (GAC) is employed to remove the excess chlorine and/or any chlorinated compounds that might still remain in the water. At the same time, the GAC will also remove any of the undesired substances that may have passed through the coagulation chlorination treatment (see Chapter 9, "Conventional Treatment Plant"). This procedure results in the most efficient use of GAC.

In some instances in which the plant water supply is of high quality, the GAC unit is used for both filtration and chlorine removal; however, such an arrangement requires close observation that no new problems might arise. These carbons are specifically designed for the removal of organic impurities found in municipal supplies. This phenomenon is termed *adsorption,* which means "to collect on a surface". Note the difference from *absorption,* which is defined as "to drink up, suck up, or to swallow." Adsorption is attained because of the unique pore structure that results from the activation process previously described.

Activated carbon has a high tenacity for taste, odor, color, phenols, tannins, and other organic matter found in water, but its capacity for such varies in its applications. Excess chlorine, permanganate, and other oxidants used in the pretreatment and disinfection of product water applications are also effectively

Figure 12-1. Activated carbon particle (stylized).
Courtesy, Westvaco Corp.

removed.

Carbon removes the various unwanted substances by various phenomena:

● The carbon acts as a filter to remove or trap those substances that have not been destroyed by chlorination. These particles are said to be *adsorbed* on the surface of the pores of the carbon. To assure adsorption, the pores must be of the approximate size of the particles under discussion, and the capture of such material will eventually result in the plugging or fouling of the pores, to reduce the effectiveness of the carbon. It is thus a wise policy to apply as few such particles to the carbon as possible.

● The removal of the chlorinated substances is accomplished by direct chemical oxidation of the carbon. This action will eventually reduce the size of the carbon particles and, in this manner, the effectiveness of the carbon is again reduced.

● The chlorination of water, whether through the use of chlorine gas, calcium hypochlorite, or sodium hypochlorite, can result in various chlorinated compounds that are removed in their contact with activated carbon.

Activated carbon acts as a catalytic agent that promotes or hastens the dissociation of those compounds that result when chlorine is added to water. These compounds can be HOCl or chloramines (NH_xCl_y).

Free Chlorine

If there is no ammonia in the water:

$$Cl_2 + H_2O \longrightarrow HCl + HOCl$$

High concentrations of chlorine can convert carbon to CO and CO_2, and this will cause the carbon to be consumed. In such a case:

$$C + 2HOCl \longrightarrow CO_2 + 2HCl$$

Chloramines

If ammonia is naturally present or is added to limit the production of trihalomethane, chloramines having the general formula (NH_xCl_y) can be produced.

pH 8.5 or greater	-	NH_2Cl (monochloramine)
pH 4.5 to 8.5	-	$NHCl_2$ (dichloramine)
pH below 4.5	-	NCl_3 (trichloramine)

Monochloramine (NH_2Cl). At the risk of repetition, the reader is reminded that chloramines can enter the system because of natural ammonia in the water to be treated. They can also occur through the addition of ammonia to the water to achieve both their greater stability in the water and also to decrease the formation of THM.

$$NH_2Cl + 2H_2O + C^* \longrightarrow NH_3^+ + H_3O^+ + Cl^- + C^*O$$
$$H_2O + 2NH_2Cl + C^*O \longrightarrow N_2 + C^* + 2H_3O + 2Cl^-$$

Dichloramine (NHCl$_2$) is formed by the following mechanism:

$$C^* + 2NHCl_2 + H_2O \longrightarrow N_2 + C^*O + 4HCl$$

The reactions that take place between chlorine and carbon are of such complexity that it becomes difficult, if not impossible, to answer the questions raised from time to time by operators:

● Why did the carbon last only 3 months or 5 months?

● How long should a recharge of activated carbon last?

● Do some carbons last longer than others, and why?

● How much chlorine should activated carbon remove?

These questions are interrelated and, for some reason, seem to be occurring more and more. These questions imply that carbon life is probably one of the least understood facets of chemistry in water treatment. A review of the literature and actual operating experience over many years leads to the definite conclusion that the best answers to these questions are empirical.

Many claims are made for the absorptive, adsorptive, and reactive powers of the various carbons, based on so-called standard tests devised by the various carbon manufacturers. Terms such as *iodine number; methylene blue number;* or *phenol, chlorine, red, green,* or *violet numbers* are given that represent the adsorptive values for carbon of these dyes or compounds. The practical value of these terms as far as their relationships to the removal of chlorine, taste, or odor is somewhat obscure.

For example, one instance is cited wherein it was stated that red-striped peppermint candy was dissolved in water. It was determined that one carbon removed the color 5 times longer than a competitive carbon. This test is significant only from the standpoint that it illustrates a particular method for removing color from a solution prepared in a specific manner. The test, however, could hardly be used as a measure to compare the merits of the carbons from the standpoint of taste and odor removal, effective life, or efficiency. Similarly, those tests using dyes such as methylene blue show like discrepancies.

If activated carbon were used to remove chlorine from distilled water at constant temperature, it would be possible to determine how much chlorine could be removed by a unit quantity of carbon. However, water treated under actual conditions contains innumerable constituents that affect the efficiency and life of the carbon. Some of these are familiar, and others are of a complex organic nature. It would be impossible to establish a set of test procedures that would truly measure that which is required.

None of these tests is a true or satisfactory measure of the dechlorinating power of a carbon. Suffice it to say that the only meaningful test that the carbon user finds of value is the length of time the carbon remains efficient under actual operating conditions. Continuing research will no doubt lead to more specific recommendations; for the present, however, the approach remains empirical.

The literature available, both from manufacturers and also by water chemists, is quite extensive. The discussion that follows is a mixed summary of the information that can be gleaned from such reading.

There seems to be agreement on the fact that high-quality carbons such as those manufactured from coal carbons and from properly processed cellulose and/or high-grade nut shells are superior carbons. Most of these are excellent for dechlorination, but some do not satisfy other criteria. For example, some of these carbons are not suitable for removal of tannins, a natural product that causes taste and odor.

The activated carbons in use today in most plants range in particle size from 20×50 to 12×20 mesh. Since different manufacturers specify different particle sizes, there will, of course, be different claims. For example, the manufacturer

Table 12-1
Properties of Various Carbons

Property	Carbon #1		Carbon #2		Carbon #3
Mesh	12×40	20×50	12×50	20×40	20×50
Iodine no.	850	850			800-900
Methylene blue no.	210	210			180-200
Ash	7%	7%	6%	6%	
Abrasion no.	70	70			25/100
Surface area, m²/g	820	820	625		
Voids space	40%	40%			
Apparent density, lb/ft³	36	36	20		30
Chlorine adsorption					400+
Pore volume, mL/gm			0.95		

Mesh - Size of carbon particles.
Iodine and methylene nos. - mg of substance removed by 1 gram of carbon.
Chlorine adsorption - This value indicates the number of gallons of water containing 20 ppm of chlorine that will pass through a 1-inch-diameter tube, 12 inches deep, before the appearance of 1 ppm chlorine in the effluent.
Abrasion - Test for the hardness of the carbon particle. The higher the value, the harder the particle.

whose carbon is finer will claim that fine carbon will remove more chlorine more efficiently. The claim continues to say that a finer carbon has more surface contact area. It is obvious that powdered carbon should give the most efficient results, and it is used in this form in municipal treatment plants. However, there is a practical limit to the fineness that can be tolerated when the carbon is used continuously in a packed column such as a carbon purifier used in a production plant.

On the other hand, the manufacturer of the coarser carbon, even though it might also manufacture the finer granular carbon, claims that the lower back-wash rate required by the fine carbon does not sufficiently scour the carbon particles, thus reducing its efficiency in time. The backwash required for some heavier granular carbon at normal temperatures is the same as that required for sand particles, and this will assure the proper scouring, scrubbing, and discharge of the accumulated material.

Table 12-1 is a listing of some of the properties of the five grades of commonly used activated carbons.

The information in the preceding section seems to be saying that the various manufacturers may not yet have devised a test that will conclusively establish the dechlorinating power of a GAC.

There seems to be no doubt that this efficiency will be determined by the contact time required between a particular water and a particular GAC. Nevertheless, all GAC manufacturers agree on the following factors as affecting dechlorination efficiency:

● The chemical type of chlorine or chlorinated compound that is to be removed.

● The pH and temperature of the water under consideration.

● The organic content of the water (some carbons have a greater capacity than others for removal of organic matter).

● Rate of flow through the carbon bed. The present design parameter is 1.0 gal/min/ft^3 of carbon.

As a result, if there are problems with any of the more or less standardized sizes of a carbon bed, the alternative would be to set up a test carbon column. With such a device the most efficient contact time can be determined by varying the flow and extrapolating into an operating unit.

The carbon will become exhausted first at the point at which the water is applied, and this exhaustion will progress downward through the bed until the entire bed will no longer remove the undesirable chlorinated compound. In practice, a warning or alarm can be signaled by sampling the water taken at different levels of the carbon bed. However, it will probably be determined that once the upper level of the carbon bed is no longer effective, not much time will pass until the complete bed becomes exhausted.

Carbon beds in series, using the first bed for roughing and the second for polishing, have been attempted; but there really is little savings in carbon, had the same total amount of carbon been used in a single unit.

Among the problems encountered in the use of carbon towers is leakage or short-circuiting across the valve in the purifier, from the piping that separates the inlet and outlet flows. One way to check this is to test a sample of water taken from the bottom of the purifier tank. If such is not possible, a sample from the tank's drain-to-waste valve should be checked. Of course, in both instances the samples should be collected while water is flowing through this unit. If these tests indicate an inlet chlorine residual and also more than a trace of residual in the effluent, check the drain-to-waste valve. If this separation valve is tightly closed, it could indicate that the carbon bed is insufficient.

If there are flow surges that exceed the capacity of the carbon purifier, and if they last long enough, there is a danger of incomplete dechlorination.

The main purpose for backwashing a carbon bed is to free the bed of the compaction that results from the downward flow of water through the bed, and to remove any gas that could be attached to particles. A second reason for backwashing is to rearrange the carbon particles so that the same carbon particles are not always at the upper portion of the bed. Thus, along with removing any particulates from the surface, a proper backwash will also scour and reposition the carbon particles (if the proper rates are used).

It is possible, at times, that certain materials that are captured by the carbon particles will be more efficiently removed following a steaming of the bed. If none of the above suggestions seem to uncover the cause, a sterilization with steam could possibly help clean the bed. A number of operators will claim that this procedure will reactivate the carbon. This is doubtful since the temperatures reached are likely to be not much more than 200 °F. However, it is possible that the steaming can help to clean the carbon particles.

Which activated carbon is preferred? The answer would seem to narrow down to several factors:

● Purchase from a reputable manufacturer whose business is the manufacture of activated carbon for water treatment.

● Purchase from a manufacturer who will be able to furnish day-to-day service to the operator.

Last but not least, consider economics. When a recharge of carbon is required, the same volume of the heavier carbon will require more pounds than does the less dense carbon to occupy the same space in the purifier. None of the manufacturers will guarantee the life of a carbon. All of them recommend that carbon be replaced each year, based on an 8-hour day, before the busy season.

For example, a 48-inch-diameter purifier using a 20-inch-deep carbon bed will require 21 ft^3 of activated carbon:

Area of 48-inch unit = 12.58 ft^3
Therefore, a 20-inch bed will require:
    12.58 ft^3 × 20 inches ÷ 12 inches = 251.6 ÷ 12 = 20.95, or 21 ft^3

If carbon weighs 20 lb/ft^3 and costs \$1.00/lb, then a recharge will cost:
    21 ft^3 × 20 lb/ft^3 × \$1.00/lb = \$420.00.

If carbon weighs 40 lb/ft^3 and costs \$0.50/lb, then a recharge will cost:
    21 ft^3 × 40 lb/ft^3 × \$0.50/lb = \$420.00.

Thus, it can be seen that although the heavier carbon costs half as much per pound, the recharge cost is the same.

Although this chapter has been concerned with the removal of chlorinated substances, note in Table 10-2 perhaps an equally important role that GAC plays in the removal of those by-products that result from disinfection.❏

CHAPTER 13

FILTRATION

Conventional (Particle-Removal) Membrane

A filter was originally defined as a device for separating insoluble solid particles from water by passing it through a porous medium. A *porous* substance is one that is full of tiny holes through which the water can pass while the solids are retained.

Today, however, with the recent development of new polymer materials, it has become possible to design filters that will remove much smaller particles than was previously possible, and others that will also remove soluble solids.

For purpose of discussion, the first group of those above will be termed conventional (particle) filters (Figure 13-1), and the latter group will be called membrane filters (Figure 13-2).

Each of the various filters is identified by the respective pore size of its medium (e.g., the smaller the pore, the smaller the particle that can be removed). Further, because of the wide range of particle sizes involved, there is a tendency to use several different units of measurement.

For example, it is noted from the filtration spectrum (Figure 13-3) and other research that at least five units of measurement are listed. These are micrometer, micron, angstrom, nanometer, and molecular-weight cutoff.

Suffice it to say that if only one unit were used, it would probably be the micrometer, also called the micron (μm). An enumeration of the various filters developed for the manufacture of product water follows, but first note several differences between the two groups of filters:

● All of the water applied to the conventional filters passes through the medium. This is not the case with some of the membrane filters. In fact, as much as 25% of the water applied to some membrane filters might be diverted as waste.

● The water that passes through a conventional filter does not undergo any change in its chemical composition, but there is a change in the chemical characteristics of the water that is filtered through some membranes. The amount of change is dependent upon the membrane under consideration.

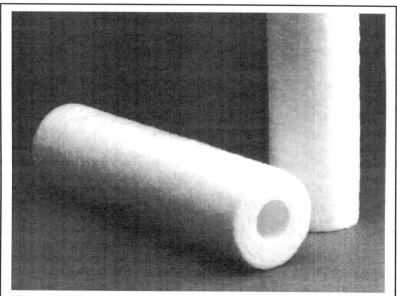

Figure 13-1. Particle filter element.
Courtesy, Osmonics Inc.

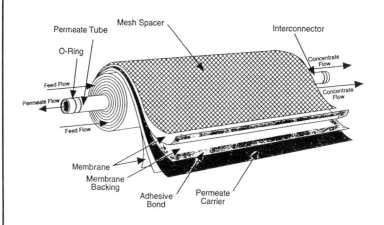

Figure 13-2. Membrane element, exploded view.
Courtesy, Osmonics Inc.

Table 13-2 gives an idea of the approximate range of pore sizes of the conventional filters.

Strainers

If a production plant obtains its water from a lake or stream, strainers in all likelihood can be beneficial. Strainers are considered to be *roughing filters* when used in water treatment to remove larger particles. A macrostrainer is capable of removing particles in the range of 100 μm or greater. A microstrainer will capture particles below 100 μm.

The classical screening mechanism can be designed either about a rotating or a fixed strainer. The rotary strainer is usually in the form of a rotating horizontal drum that is covered by a screen whose mesh size or slot size is selected to fit the requirements. The raw water usually enters the inside of the drum and flows outward. Jets of water are used to keep the screens clear and also to help the discharge of the filtered particles from the system.

Typical in concept, but not necessarily in design, a relatively newer type of in-line strainer (see Figure 13-4) seems to be gaining in application. This equipment is used to remove substances such as dirt particles, algae, rust, and sand from the water. Even though it may not completely remove all of the unwanted particles, it can still be used prior to further treatment of the water.

Polishing Filter

Practically all conventional treatment plants include a filter to capture any solids that might have passed through or from the sand filter or activated carbon purifier. In the trade, such a unit is called a *polishing filter,* and it consists of

Table 13-1
Units of Measurement

1.0 micron (μm) = 1.0 micrometer = 1×10^{-6} meter
          = 1/1,000,000 meter
          = 1/25,000 inches
          = 10,000 (10^4) angstrom units

1.0 angstrom   = 1/100,000,000 centimeter
          = 1×10^{-10} meters

1.0 nanometer = 1×10^{-9} meters
          = 1/1,000,000,000 meter

To give some perspective:
Diameter of human hair = 75 to 90 μm
Grain of salt = 90 to 100 μm

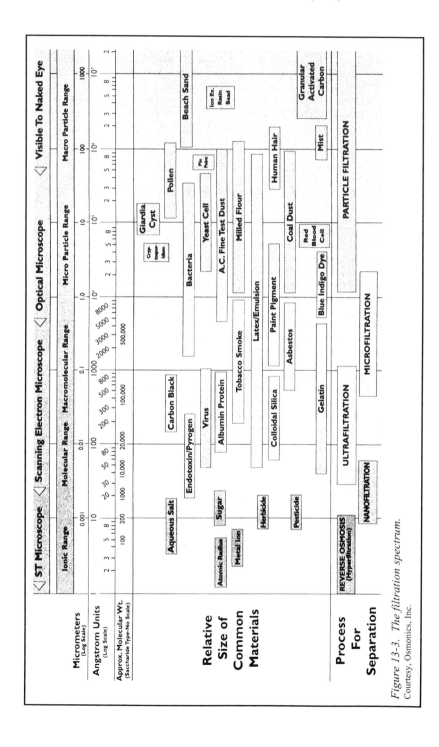

Figure 13-3. The filtration spectrum.
Courtesy, Osmonics, Inc.

disposable cartridges of various types that are housed in stainless steel containers. Regardless of the manufacturer, the cartridges are assembled as shown in Figure 13-5. The cartridges are manufactured by spirally winding a string, yarn, or twine about a rigid, perforated core. The tension in the winding is controlled to deliver a firm and homogenous medium.

Cartridges are made in 10-, 20-, 30-, and 40-inch lengths to furnish a wide choice of flow ranges. The materials used to make up the cartridges are bleached white cotton and/or propylene. If these are to filter potable water, they must be so labeled.

If these polishers will be receiving dechlorinated water, bacterial buildup can become a problem; and the cotton and polypropylene can both withstand the required 100 ppm of chloride needed for sterilization.

The life of the cartridge will depend upon the quality of water to which it is applied. In a normal operation (when there is pretreatment), these cartridges will last about 4 weeks when filtering at a rate of 3 gpm for each 10-inch length of cartridge. The initial loss of head is in the order of 3.0 pounds per square inch gauge (psig). The cartridges are not disposed of until there is a noticeable drop in throughput flow. If a cartridge is used too long, it will be found that solids will have penetrated through the entire thickness of the cartridge. Figure 13-6 shows how this penetration can occur.

While most applications will use a 10- to 20-μm filter, cartridges can be obtained to remove particles as small as 5 μm. To visualize the effect of various sizes of particles, note the following from a bulletin issued by a filter manufacturer:

"To remove visible specks a 40- to 50-μm filter should be specified. To produce optical clarity in a liquid, a 25-μm or finer is required. To remove a haze from a liquid, a 10-μm or finer filter should be used."

Diatomaceous Earth
Diatomite is described in an older Johns-Manville *Bulletin Fa-84A581* as follows:

"Celite Diatomite is the skeletal remains of tiny aquatic plants called diatoms. They flourished in prehistoric waters over what is now Lompoc, California. Over the eons these skeletons formed a deposit on the ocean floor which then rose to become a part of the land mass. The celite deposit is distinguished by high purity and an almost infinite variety of diatom shapes and sizes."

Diatomaceous earth (DE) has been widely used as product filters in industries such as sugar, oil, and beverage for many years, and it is now finding use in municipal water filtration for general use and also for removal of the *giardia* cyst. It is also a popular filter for use in swimming pools. These filters are of the pressure or vacuum type, which consist of septa made up of a core that can be either flexible or rigid.

One particular element uses a flexible core, as illustrated in Figure 13-7. Once the precoat has been applied, the flow of water is through the precoat to the interior of the filter tube or element and out to process.

Diatomaceous earth is prepared in a number of grades. It is normally a white or pink powder with a density of 7 1/2 lb/ft^3 and is capable of removing particles in the order of less than 0.5 μm.

The following (Coopermatics,) is the description of a DE filter used in the soft drink industry. This particular unit (Figure 13-8), which uses a filter cake composed of a mixture of DE and powdered activated carbon, is capable of removing not only small particles but also colloidal iron and the residual free and combined chlorine (less than 0.5 $\pm$ ppm residual) found in the municipal supply.

Purification is accomplished by forcing the Raw Water through a Filter Cake which retains the dirt particles and removes chlorine as the raw water passes through. The Filter Cake is made up of D.E. and powdered carbon properly supported by the filter tubes. The filter tubes must first be uniformly coated with D.E. and carbon. The control system automatically provides a "Precoat Phase" to deposit the D.E. and carbon on the filter tubes before bringing the system to its "On-Stream Phase."

As a porous Filter Cake picks up dirt particles, more and more pores become blocked until the flow rate of the filter starts to decrease. A new set of pores can be made available by breaking up the Filter Cake with a "Bump Phase" and redepositing the Filter Cake on the filter tubes with a "Precoat Phase". The dirt particles which were coating the surface of the Filter Cake are now mixed rather uniformly all through the Filter Cake, leaving the surface relatively clean. The "Bump" and "Precoat Phase" can be repeated many times to "Regenerate" the Filter Cake. When the Filter Cake is removed from the filter by the "Transfer Phase" which is started and stopped by the operator.

Table 13-2
Pore Sizes for Conventional Filtration

Particle Filtration	Range of Pore Size
Macrostrainer	>100 μm
Microstrainer	<100 μm
Coagulation, settling, & coarse media filter	several μm
Sand or coal	
Polishing filter	5-10 μm
Diatomaceous earth	0.5-200 μm

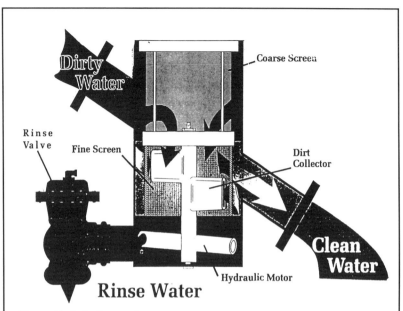

Figure 13-4. In-line strainer.
Courtesy, Filtomat

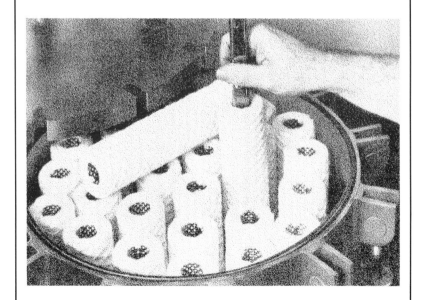

Figure 13-5. Cartridge filter.
Courtesy, Cuno Inc.

The spent Filter Cake can be discharged to sewer depending on the locality. If this method of spent Filter Cake disposal is not satisfactory the Coopermatics System will include a Sludge Receiver which is capable of accepting the entire contents of the filter. At the end of a filter cycle the contents of the filter are transferred to the Sludge Receiver and the filter is immediately returned to the "On-Stream Phase" after being charged with D.E. and carbon. The Sludge Receiver is placed in the "Blow-Down Phase" so that the volume of slurry it received from the filter will be deliquefied and the sludge dumped as a Semi-Dry Cake. Once the Sludge Receiver has been emptied it will be ready to receive another load of slurry from the filter.

Depending on the physical quality and free chlorine content of the raw water, it appears that a spent Cake of 75 pounds, dry solids, will result when treating 150 GPM for 40 to 80 hours of operation.

Membrane Filtration
The membranes under discussion are classified as *semipermeable,* which indicates that they will allow the passage of water through the membrane pores while retaining various amounts of suspended and/or dissolved substances. Before proceeding to any further discussion, it would be wise to review some of the terms employed in membrane technology:

● The raw water to be treated is called *feed.*

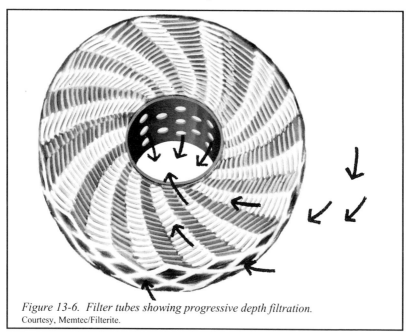

Figure 13-6. Filter tubes showing progressive depth filtration.
Courtesy, Memtec/Filterite.

● The effluent is called the *permeate*.

● The waste from the process is called the *concentrate* or *reject*.

● The rate of flow (gpm/ft^2) is called the *flux*.

The following information is taken from a paper delivered by Gash (1986):

> A wide variety of membrane polymer types and physical arrangements for crossflow filtration are available to meet a variety of industry demands. Overall, man-made membranes comprise a plastic material that must perform in a variety of environments with high efficiency. Operating conditions such as temperature, pressure, solution pH and chemical compatibility must all be considered. The specific process objective - water softening, production of 18-megohm/cm water, TOC or particulate removal, etc. - is also a primary criterion for selection of a membrane type.
>
> No one polymer will withstand all possible operating environments nor meet all water-purification objectives. For the treatment needs of most steam-generation (for power) facilities, one of three or four standard RO and UF membranes will handle the vast majority of processing needs. The most widely used RO membranes are the CA

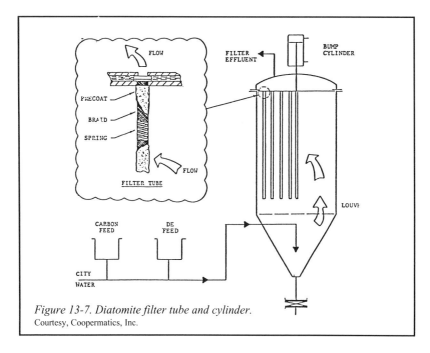

Figure 13-7. Diatomite filter tube and cylinder.
Courtesy, Coopermatics, Inc.

(cellulosic) and PA (polyamide) types, rated at 97% to 99+% NaCl rejection and 150 - 200 MWCO (molecular-weight cutoff); the most popular UF membranes are the small-pored PS (polysulfone), VF (a propriety fluorinated material), and CA types having a 1000 - 20,000 MWCO.

UF membranes are usually operated at 15 to 200 psig, and RO membranes from 200 to 1000 psig. In the pressure ranges that UF and RO are subjected to, these plastic materials must resist mechanical compression (compaction) that would deform their morphology and alter their performance characteristics. The significance of compaction has diminished in recent years with the development of RO membranes that are inherently more resistant to compaction and capable of higher flux at moderately low operating pressures (200 - 400 psig). Recent studies also indicate that proper prefiltration of feed water ahead of crossflow filtration systems significantly alleviates the irreversible fouling that was once confused with compaction.

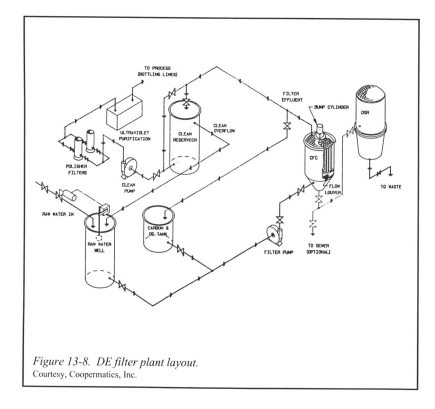

Figure 13-8. DE filter plant layout.
Courtesy, Coopermatics, Inc.

Filtration membranes are manufactured in three different configurations, each supposedly superior to the others for specific applications. Each of those that will be discussed here are used for microfiltration (MF), ultrafiltration (UF), nanofiltration (NF), and reverse osmosis (RO), but there could be some slight modification in the structure of the membrane material when applied to MF and UF.

Spiral-wound configurations (Figure 13-9) are sheet-type membranes that are encased in a container. These membranes are wound around a permeate (effluent) tube. The feedwater (water to be treated), which is under pressure, circulates through the membranes to the permeate tube and to product water use. The solids, both soluble and insoluble, that have been removed from the water remain on the outside of the membrane and are discharged through the concentrate (waste) outlet opening.

Hollow-fiber configuration (Figure 13-10) consists of a large bundle of very fine hollow fibers, each of which has a diameter of about 100 μm. These fibers are handled differently by different manufacturers, but basically the fibers are encased in a container, and the flow of water is from the outside of the fiber into its hollow core and out to process. The waste is rejected as shown.

The two configurations described above seem to be by far the most popular and, as a result, further discussion will concern only these membranes. The membranes are modular, and these modules, manufactured by different suppliers, are of different dimensions. Each module can be installed in its own tubular vessel, or a number of them can be placed in a longer tubular vessel, as indicated in Figure 13-11.

Since there is a variation in the size of membrane modules, it is not possible to discuss a standard dimension, but the information in Table 13-3 will give some idea as to the relationship between flow, recovery, and overall dimension of a complete RO purification water machine.

In previous chapters in which equipment was discussed, accepted design parameters were specified and the appropriate calculations were carried out. This procedure will not be repeated here because this particular engineering is yet somewhat unfamiliar to most operators and even to many engineers. The caveat is issued that the selection of both the membrane configuration and its application is best left to the manufacturer or consultant because of expertise and warranties.

Product Water Requirements
A wide choice of membranes — as to both configuration and pore size — is made available to the user because of the wide range of required product waters.

For example, in the soft drink industry, the limits shown in Table 13-4 are set on the water to be used.

On the other hand, the product waters required in semiconductor, pharmaceutical, and other manufacturing processes have the most rigid requirements. The ultimate goal for quality control in each of these industries is to produce a product water that contains nothing but water. This water must be of such quality that it is described as *high purity*. By definition then it must contain a minimum or even an absence of all particles: soluble or insoluble; chemical, physical, or bacteriological.

Water of this quality requires a parameter not mentioned previously. This unit of measurement, resistivity, is shown in Table 13-5 in a partial listing from a semiconductor standard.

The only parameter in Table 13-5 that may require explanation is *resistivity*: Pure water is not a good conductor of electricity, and dissolved ions will increase the conductivity of water. Conductivity will also increase directly with an increase in temperature. Consequently, the temperature standard selected is 25 °C (77 °F).

The unit of conductivity is the *micromho* (mho), and pure water will have a conductivity in the order of 0.00005 mho/cm, a rather awkward number to handle. Because of the desire to use a more convenient unit of measurement, the reciprocal of the mho has been selected. The unit is called an *ohm*, the unit of resistance.

$$1.0/\text{mho} = \text{ohm}$$
= The resistance of a circuit in which 1 volt maintains a current of 1 ampere
1.0 mho/0.00005 = 20,000 ohms

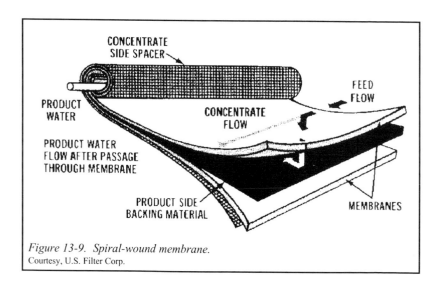

Figure 13-9. Spiral-wound membrane.
Courtesy, U.S. Filter Corp.

Membrane Characteristics

Pore sizes can be determined from the Filtration Spectrum (Figure 13-3). To convert micrometers (micron, [μm]) to nanometers (nm), multiply number of μm by 1,000.

Microfilters (range, 0.1 to 2.0 μm) will remove small suspended solids, large colloids, some emulsions, and high volumes of bacteria. They are not used to remove soluble solids or to change the chemical characteristics of the water. Microfilter operating pressures range from 1 to 25 psig. Microfilters are sometimes used as a pretreatment in high-purity water applications.

Ultrafilters (range, 0.008 to 0.1 μm) will remove large organics (over 1,000 MW) such as proteins, pyrogens, bacteria, and colloids. They act also as pretreatment and posttreatment to ion exchange and other high-purity processes. They do not remove dissolved solids and so do not change the chemical characteristics of the water. Operating pressures range from 10 to 200 psig.

Nanofilters (range, 0.001 to 0.01 μm) are a newer type of membranes. They are used to remove organics and dissolved salts, thus changing the chemical characteristics of the water. Operating pressure ranges from about 50 to 300 psig.

Reverse osmosis (range, up to 0.001 μm) will afford the greatest removal of soluble solids. Operating pressure varies from 400 to 800 psig.

By definition *osmosis* is the diffusion (passage) of a water of less concentration of dissolved solids through a semipermeable membrane to a water of greater

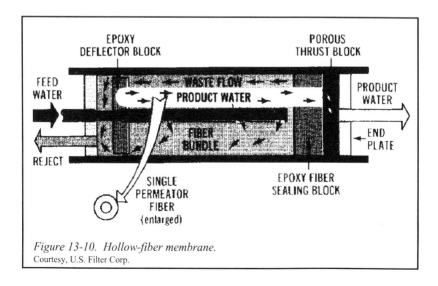

Figure 13-10. Hollow-fiber membrane.
Courtesy, U.S. Filter Corp.

concentration of dissolved solids. Perhaps this process may be more easily visualized by considering the case of a raisin that contains a high concentration of sugars.

If the raisin is placed in pure water long enough, it will swell. What has happened, of course, is that some water has passed or migrated through the skin of the raisin (a semipermeable membrane) to the inside of the raisin. The natural force that drove the water through the skin is called *osmotic pressure,* and the process is called *osmosis.* The transient water will remain inside the raisin until an outside source such as pressure, heat, and/or evaporation drives it out. The phenomenon of osmosis is shown in Figures 13-12 to 13-14.

Osmotic pressure drives the pure water from B to A. The level of water will rise in compartment A until the pressure of this column of water stops the flow of pure waters, as shown in Figure 13-13. Osmotic balance has been reached, and the value of this hydrostatic pressure is the osmotic pressure of solution A. Osmotic pressure is a physical characteristic linked to the concentration of a solution, and it increases with concentration.

The osmotic pressure of natural water is approximately 10 psig for each 1,000 ppm of total dissolved solids. Seawater, which has a concentration of salts in the neighborhood of 40,000 ppm, will have an approximate osmotic pressure of 40,000/1,000 × 10, or 400 psig.

If a pressure higher than the osmotic pressure is applied on the more concentrated solution, osmosis is reversible, and this process, as shown in Figure 13-14, is termed *reverse osmosis.* When a pressure (in actual practice 400 to 600 psig) is applied:

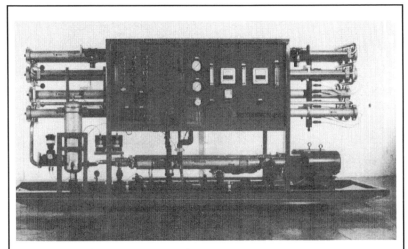

Figure 13-11. Reverse osmosis stack.
Courtesy, Osmonics, Inc.

● The natural tendency of the pure water to pass through the membrane from B to A, is of course, stopped.

● The direction of the flow is now away from the applied force so that pure water flows from A to B.

The membrane retains most substances such as dissolved mineral salts, microorganisms, and bacteria, which means that reverse osmosis can be used to purify the water in compartment A, as in the system shown in Figure 13-15.

The efficiency of removal in a single-pass RO system is in the order of 65% to 95% or higher. (More information will be given later in this chapter.) The following criteria will determine the selection of the membrane:

● Its capability for removing unwanted salts, and the efficiency with which this removal is performed.

The most important parameter in this regard is the effective feed pressure for which the membrane is designed. Note from the curve in Figure 13-16 that the salt passage through the membrane decreases, and that the efficiency of removal increases as the feed pressure increases.

● Since filter rates of RO membranes are extremely low in comparison to conventional (particle) filters, this parameter has a great influence on operation and costs.

The factor that has the greatest influence on flowrates, other than the normal resistance to flow, is that of temperature. Note, from the curve in Figure 13-17, the effect of water temperature on the flow through the membrane. This is because the viscosity of the water decreases as its temperature increases.

The productivity of a membrane is based on a feedwater temperature of 25 °C (77 °F). The flow through the membrane will increase approximately 3% per C°, or about 1.66% per F°. Although the temperature has little effect on the quality of the finished water, all guarantees are based on the type of dissolved solids to be removed, the working pressure, and the water temperature.

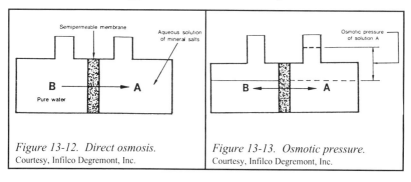

Figure 13-12. Direct osmosis.
Courtesy, Infilco Degremont, Inc.

Figure 13-13. Osmotic pressure.
Courtesy, Infilco Degremont, Inc.

Table 13-3
Reverse Osmosis Plant Information

Recovery (%)	Permeate Rate (gpd)	(m³/d)	Pump Hp (kW)	Height (in cm)	Width (in cm)	Depth (in cm)	Net lb (kg)	Shipping lb (kg)
Rejection Permeate Quality								
60/75	4,030	(15.3)	5 (3.7)	72 (183)	92 (234)	34 (86)	423 (192)	662 (301)
60/75	8,060	(30.5)	7.5 (5.6)	72 (183)	92 (234)	34 (86)	489 (222)	728 (331)
60/75	12,100	(45.9)	10 (7.5)	72 (183)	92 (234)	34 (86)	555 (252)	794 (360)
60/75	16,100	(61.0)	10 (7.5)	72 (183)	92 (234)	34 (86)	621 (282)	860 (390)
60/75	20,800	(78.8)	10 (7.5)	72 (183)	131 (333)	34 (86)	996 (453)	1,174 (534)
60/75	31,800	(120.5)	20 (14.9)	72 (183)	172 (437)	34 (86)	1,225 (557)	1,465 (666)
60/75	41,200	(156.1)	20 (14.9)	72 (183)	172 (437)	34 (86)	1,420 (645)	1,570 (714)
60/75	51,700	(195.9)	25 (18.7)	72 (183)	172 (437)	34 (86)	1,680 (764)	1,800 (818)
60/75	63,000	(238.8)	30 (22.4)	72 (183)	154 391	54 (137)	2,100 (953)	2,350 (1,066)
60/75	70,000	(265.3)	40 (29.8)	72 (183)	154 391	54 (137)	2,300 (1,045)	2,550 (1,159)
60/75	95,000	(360.1)	50 (37.3)	72 (183)	154 391	54 (137)	2,550 (1,159)	2,800 (1,273)
60/75	120,000	(454.8)	60 (44.7)	72 (183)	154 391	54 (137)	2,800 (1,273)	3,050 (1,386)

● The chemical characteristics of feedwater have a bearing on the ultimate efficiency of a membrane. For example, note in Table 13-6 the rejection rates of various ions. The rejection rate or holdback of colloids, bacteria, viruses, and almost all other organic matter is probably about 100%, but these substances can cause serious problems in the operation as will be explained later in the "Pretreatment" section.

● The stability of a membrane is a very important consideration, as the cost and life of a membrane make up a large part of a plant's cost. Module life is usually guaranteed for some 3 years. If for some reason the module needs replacement before that time, the cost is prorated.

Reverse osmosis reject water. Another item that requires discussion is the quantity and quality of the reject water. It is quite possible that the disposal of this highly concentrated, highly mineralized waste can be a difficult problem.

Consider, if you will, that in normal practice, for each 100 gallons of raw water that is sent to the unit, there will be 75 gallons of treated water. This means that 25 gallons will be the reject that is sent to waste. Therefore, the salts removed from the water would be concentrated (100/100) - 75 = 4 concentrations.

Assume that the total dissolved solids in the raw water equals 1,000 ppm. This is not unusual in a water that is to be treated by reverse osmosis. The concentration of the reject can be as high as 4 x 1,000 = 4,000 ppm (mg/L), which is 0.4% (assuming 100% removal).

Now assume that a plant would require 72,000 gallons per 8-hour day (150 gpm). This would mean that to get 72,000 gpd it would be necessary to feed 72,000/0.75 = 96,000 gpd to the unit; and the reject would amount to 96,000 - 72,000 = 24,000 gpd.

The volume of such waste might be reduced by passing it through a second membrane (Figure 13-18), but of course the concentration of the reject from the

Table 13-4
Sample Water Limits for RO in Soft Drink Industry

Parameter	ppm
Iron	0.1
Hardness	200
Alkalinity	50-75
Sodium	Varies
Chloride	250
Sulfates	250
Total dissolved solids	500

(There are of course limitations on suspended solids and color.)

second volume will have increased considerably.

Pretreatment

Because the pores or "water passage" through which the water flows during treatment are extremely small (less than 0.002 μm), it becomes imperative that no substance that will clog these openings is contained in the water applied to the system. Fouling of the membranes, whether caused by inorganic or organic particulate or scaling, will impede and decrease the flow through the RO membranes, and as a result will increase the passage of salts, which in turn will decrease the quality of the finished water.

The items that are monitored closely in order to keep a close check on an RO system are flow, salt passage, and drop in pressure through the system. Any change or combination of changes in the design parameters will cause fouling of the membrane. Even under the best circumstances, it seems, some of this membrane contamination will occur. As a result, a cleaning process such as the use of hydrochloric acid, shock chlorination, or caustic is involved, depending on the nature of the problem.

Several contaminating substances cause the most concern:

Colloidal particles, inert or organic, are obviously detrimental to operation. These particulates along with iron and/or manganese will decrease efficiency. Most manufacturers request a maximum of no more than 0.05 ppm, although some membranes can tolerate more iron under low pH conditions and low oxygen content. Such conditions will keep these elements in solution.

When aluminum sulfate is used to pretreat the water (a common municipal treatment practice), and then the pH is lowered (usually the first step in the RO process), there is then the possibility that the resulting aluminum (Al_2O_3) will precipitate out, causing floc problems.

There are various pretreatments that can be used to remove particulates. First,

Table 13-5
Some Semiconductor Water Standards*

Resistivity (at 25 °C [77 °F])	17-18 megohm-cm
Particles (<0.1 micron)	2
S_1O_2	3-5 ppb
Dissolved O_2	0
Total dissolved solids	<1.0 ppm
Residue on evaporation	<10 ppb

*These values are used only for purposes of illustration. The reader is advised to check for up-to-date standards in any particular industry.

the amount and type of particulates are measured by filtering a sample of the raw water through a 0.45-µm filter and noting the speed with which the filter plugs. This gives the silt density index (SDI), and each different membrane will have its own requirements (see Appendix).

Keep in mind that all RO units are equipped with a 5- to 10-µm prefilter already. Unless the water applied to this prefilter is substantially free of particulates, however, problems loom from the start. It can be said that pretreatment for pretreatment is required.

Depending upon the water applied to the RO unit, one or more of the following pretreatments can be required:

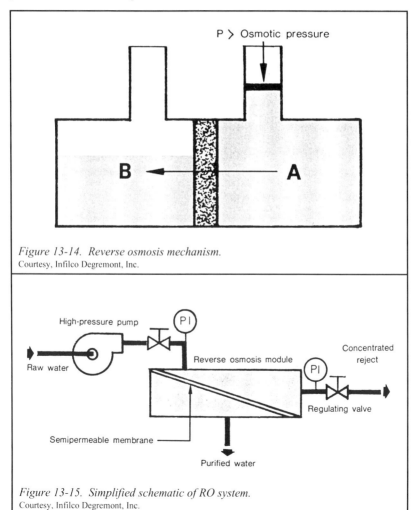

Figure 13-14. Reverse osmosis mechanism.
Courtesy, Infilco Degremont, Inc.

Figure 13-15. Simplified schematic of RO system.
Courtesy, Infilco Degremont, Inc.

1. Inert particulates can be removed by pressure filtration, whether the media is sand, anthrafil, or diatomaceous earth.

2. It is possible that coagulation and sedimentation preceding the filtration are necessary for effective particulate removal.

3. The particulates that result from the addition of potassium permanganate ($KMnO_4$) and sodium hydroxide (NaOH) will oxidize the iron and manganese so that these can be removed in a manganese greensand filter.

4. In the case of the re-precipitation of the aluminum hydroxide ($Al_2(OH)_3$) floc mentioned previously, a sand filter or the like will have to be used.

Organic matter in the form of bacteria or algae is usually found in water supplies. Such must be removed to eliminate the possibility of slime formation. As a matter of fact, some membranes are built of material that can serve as food for bacteria, and the resulting slime can and will damage this type of membrane.

If the particulate membrane can withstand chlorine, then chlorine can be added to the water prior to the RO unit. In such a case, continuous chlorination to whatever extent the membrane can withstand will allay or remove this problem.

However, some membranes cannot tolerate any chlorine. In such a situation, it might be required to resort to the conventional treatment of coagulation, filtration, and activated carbon. Not only will the algae, bacteria, and chlorine be removed, but so will the iron and manganese along with much of the alkalinity

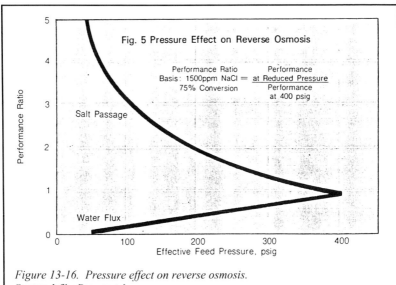

Figure 13-16. Pressure effect on reverse osmosis.
Courtesy, Infilco Degremont, Inc.

and some calcium and magnesium ions.

A second option would be to chlorinate, followed by dechlorination, as with sodium bisulfite. Even with this, it is still possible and actually preferred to arrange for a clean-in-place system for occasional cleaning.

Scale. The most likely application for RO will be to treat those waters that are apt to cause the formation of scale on a membrane surface. These scales are usually caused by the calcium salts of carbonates, sulfates, and perhaps fluorides. Scaling by these salts is by no means rare. The treated water furnished by most municipalities can scale up the insides of water pipes in homes and industries to the degree that the flow through them can be severely restricted.

To overcome this problem when using RO, sulfuric acid (H_2SO_4) is added to the incoming water to remove some bicarbonates.

$$Ca(HCO_3)_2 + H_2SO_4 \longrightarrow CaSO_4 + 2CO_2 + 2H_2O$$

The amount of this acid addition is based on the Langelier saturation index (LSI), which determines that point at which the water is neither scale-producing nor corrosive.

Typical Performances
Table 13-7 shows details of an RO plant used to reduce the total dissolved solids from 786 ppm to below 500 ppm, which is the permissible limit of TDS for

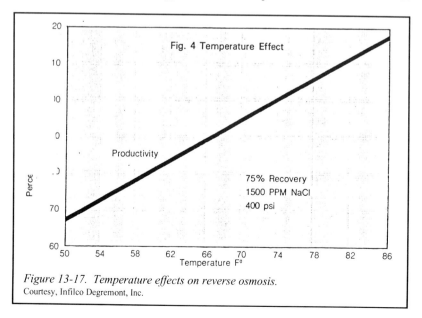

Figure 13-17. Temperature effects on reverse osmosis.
Courtesy, Infilco Degremont, Inc.

beverages, and which brings sodium down to limits needed for low-sodium drinks.

To convert calcium as ions to calcium as $CaCO_3$, multiply calcium as ions by the equivalent weight of calcium:

$$1.3 \times 100/40 = 1.3 \times 2.5 = 3.2$$

To convert sodium:

$$60.1 \times 100/2 \times 23 = 60.1 \times 2.17 = 130.4$$

On an RO unit, there are few functions to be performed by the operator, whose task is mostly observation. Figure 13-19 shows a system diagram for treatment of seawater.

Electrodialysis (ED)

Perhaps the simplest definition of dialysis is:

"... the act or process of separating mixed substances in solution, as crystalloids and colloids, by means of a moist membrane that crystalloids will pass easily and the colloids slowly if at all".

Yet, as simply as this is stated, several of the terms require further explanation. *Mixed substance* refers to the ions such as calcium, magnesium, sodium,

Table 13-6
Ion Rejection Rate by Reverse Osmosis Membranes

Ion	Low-pressure membrane (%)	Moderate-pressure membrane (%)	High-pressure membrane (%)
Ca^{2+}	85-90	95-98	99.4
Mg^{2+}	85-90	95-98	99.4
Na^+	50-60	90-95	96.5
HCO_4^-	50-60 pH dependent	70-90	pH dependent
SO_4^{2-}	85-90	95-98	99.4
Cl^-	50-60	90-95	98.5
NO_3^-	50-60	80-90	97.5
PO_4^{3-}	8-90	95-98	99.7
Pressure ranges (psig)	180-250 psig	400-500 psig	800-1,000
Applications	Softening, low TDS water <1,000 ppm	Brackish water, potable & process	Seawater, highly brackish water

bicarbonates, hydroxides, and chlorides that exist in the water supply to the manufacturing plant.

Recall that if common salt, NaCl, is dissolved in water, this dissolving process causes the NaCl to break down into Na^+ and Cl^-. This same process occurs when any compound is dissolved in water (e.g., calcium sulfate $[CaSO_4]$ will disassociate into Ca^{2+} and SO_4^{2-}). Thus, the water to undergo treatment through dialysis contains the positive and negative ions that are to be removed.

There are various types of dialysis:

Simple dialysis, like osmosis, is a natural force, and like osmosis is effected when there is a difference in concentration on either side of the membrane.

Table 13-7
Reverse Osmosis Performance

Average membrane TDS removal	82%-90%
Feed flowrate	227 gpm (51.5 m³/h)
Concentrate flowrate	57 gpm (12.9 m³/h)
Permeate flowrate	170 gpm (38.6 m³/h)
Specified recovery	75%
TDS removal at specified recovery	83%
100 - (permeate TDS/feed TDS x 100)	
Specified operating temperature	60 °F (15.6 °C)
Removal of organics over 400 MW	>99%
Specified operating pressure	430 psig (3,004 kPa)

Chemical Analyses

	Expected feed (mg/L as CaCO₃)	*Expected permeate (mg/L as CaCO₃)*	*Expected permeate (mg/L as CaCO₃)*
pH	8.0	5.4	
TDS	786	135	
Hardness	71	4	
Calcium	52.0	3.1	1.3
Magnesium	20.0	1.1	0.3
Sodium	715.0	131.1	60.1
Potassium	0.0	0.0	0.0
Alkalinity	110.7	24.1	29.3
Sulfate	327.3	19.7	19.0
Chloride	348.0	91.6	64.9
Nitrate	0.0	0.0	0.0
Fluoride	0.0	0.0	0.0

Piezodialysis is obtained when the separation of solids and water takes place with the application of pressure. In this respect, this can be likened to reverse osmosis, and lastly,

Electrodialysis is the term applied to the process when the motivating force is direct current (DC) electricity.

The logical question that arises is, "What's the difference between osmosis and dialysis?" The difference in the processes, except for the forces applied, is in the membranes.

With a semipermeable membrane, as shown in Figure 13-20, the water passes through the membrane, and the solids are trapped. This is reverse *osmosis*. With a membrane permeable to ions, as shown in Figure 13-21, the ions pass through

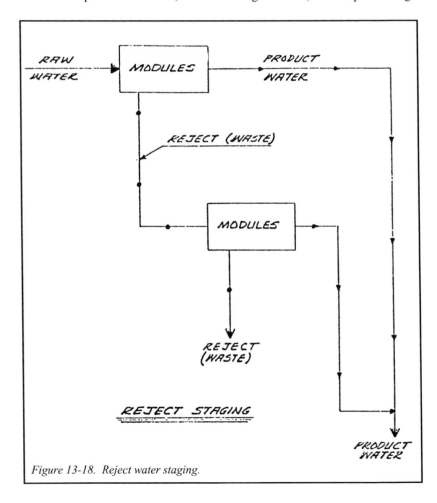

Figure 13-18. Reject water staging.

the membrane while the water is passed through. This is electro*dialysis.*

This passage of ions through the membrane is called *ion migration,* and this can be explained as follows:

For water purification by ED, the force applied is a DC current. This current flows in one direction, is steady, and is free of pulsation. (The other energy required is the pumping cost to get the water through the system.)

Direct current is the type of electricity that is obtained, for example, from an automobile battery. On the battery there are two connections (poles). One is positively charged and is called the *cathode,* while the other is negatively charged and is called the *anode.* If these two poles are connected directly to one of each of two membranes, one membrane will carry a positive charge, and the other a negative charge.

Positive ions are rejected by the positively charged membrane, but are attracted by the negatively charged membrane. Negative ions are rejected by the negatively charged membrane but are attracted by the positively charged membrane. (See Figures 13-22 and 13-23).

Figure 13-24 illustrates a single cell of an EDR system, and in actual practice a number of such cells are stacked together to make a complete EDR system. (See Figure 13-25.)

Operation. In a commericial process, the electical polarity on the membranes is reversed periodically to minimize scaling and bring recoveries in the range of 80% to 90%. This is referred to as electrodialysis reversal (EDR). The water passage in a typical EDR unit is in the order of 40 mils (approximately 0.04 inches), and as a result, only approximately 90 psig is required to push the water through the system. This is in contrast to the 400 to 800 psig required in a typical RO unit. Since the flow in the EDR is not through the membrane but over the surface of the membrane, the pumping costs will be much less than that required for the RO process. In addition to the pumping requirements, there is a cost for

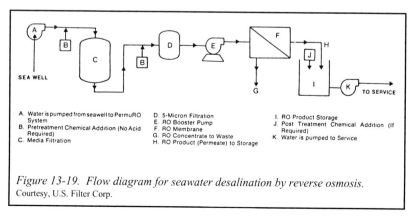

SEA WELL

TO SERVICE

A. Water is pumped from seawell to PermuRO System
B. Pretreatment Chemical Addition (No Acid Required)
C. Media Filtration

D. 5-Micron Filtration
E. RO Booster Pump
F. RO Membrane
G. RO Concentrate to Waste
H. RO Product (Permeate) to Storage

I. RO Product Storage
J. Post Treatment Chemical Addition (If Required)
K. Water is pumped to Service

Figure 13-19. Flow diagram for seawater desalination by reverse osmosis.
Courtesy, U.S. Filter Corp.

the DC current. While the pumping cost remains constant, the cost of the electricity will vary in direct proportion to the total dissolved solids of the water supply.

Electrodialysis will result in a 60% to 85% removal of the dissolved solids, depending upon the dissolved solids in the feedwater, temperature of the feedwater, and the number of stages used. The volume of waste from this process will amount to 20% to 25% of the volume of water applied.

The operation and performance of ED is quite similar to that of RO in that similar pretreatments may be required. The pretreatments need to bring the water to within the following parameters: zero free chlorine, less than 0.3 ppm each of hydrogen sulfide and iron, less than 0.1 ppm of manganese, and less than 2.0 Jackson turbidity units (JTU).

As in the case of the other membrane processes that have been discussed, the reader is strongly advised to consult with the manufacturer when designing such a system. Laboratory and even pilot-plant work may have to be performed to

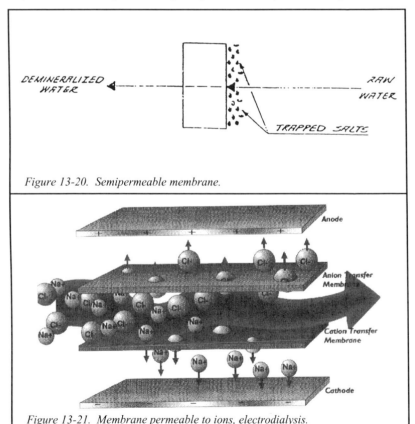

Figure 13-20. Semipermeable membrane.

Figure 13-21. Membrane permeable to ions, electrodialysis.

Table 13-8
Operating Specifications, Aquamite XX (Ionics, Inc.)

Feedwater TDS	1,500 ppm			2,500 ppm			3,500 ppm		
Number of stages	2	3	4	2	3	4	2	3	4
Temperature: 60 °F									
Product flow (1,000 U.S. gpd)	264	264	264	240	240	240	210	210	210
Product TDS (ppm)	350	185	95	630	350	180	950	500	250
% TDS removal	77	88	93	75	86	93	73	86	93
% water clarity (product to feed)	80	78	76	72	70	68	66	64	62
Electric consumption (kWh/1,000 U.S. gal)	7	7	7	10	10	10	12	13	13
Temperature: 80 °F									
Product flow (1,000 U.S. gpd)	264	264	264	264	264	264	240	240	240
Product TDS (ppm)	255	125	60	475	250	120	750	350	160
% TDS removal	83	92	96	81	90	95	79	90	95
% water clarity (product to feed)	78	76	74	72	71	68	64	63	61
Electric consumption (kWh/1,000 U.S. gal)	7	7	7	9	9	10	13	14	14
Temperature: 100 °F									
Product flow (1,000 U.S. gpd)	264	264	264	264	264	264	264	264	264
Product TDS (ppm)	255	120	60	375	160	90	770	350	130
% TDS removal	83	92	96	82	94	96	78	90	96
% water clarity (product to feed)	78	76	74	72	70	68	65	63	61
Electric consumption (kWh/1,000 U.S. gal)	7	7	7	9	10	10	12	13	13

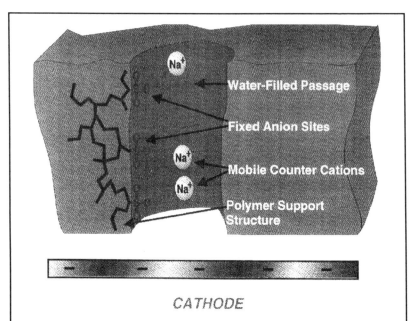

Figure 13-22. Simplified schematic of electrodialysis process, cation exchange.
Courtesy, Ionics, Inc.

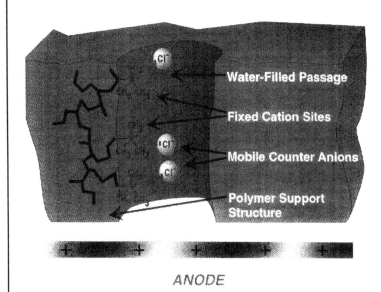

Figure 13-22. Simplified schematic of electrodialysis process, anion exchange.
Courtesy, Ionics, Inc.

design the proper unit for any particular water.

One problem common to all membrane usage is that of "dirty" or fouled membranes. The cleaning can involve some rather extensive programs.

In order to overcome such cleaning procedure problems, Ionics, Inc. (Watertown, Mass.) has developed its electrodialysis reversal (EDR) system, as described in its bulletin TP-306:

Cleaning is done automatically every 15 minutes during the operation of the plant by reversing the polarity. When the DC electric power is reversed, the anode becomes the cathode and vice versa. At the same time, the hydraulic flow is reversed, in that the product stream becomes the concentrate stream and vice versa. By reversing the polarity, the electrically charged particles (ions) are driven the opposite way from the previous polarity; thus scale or other particles which may have started to build up on the membranes will be purged back into the waste stream. This polarity reversal eliminates the need for continuous feed of acid, polyphosphates or, in many cases, the need to pretreat the feed water. This feature has significantly increased the ease of operation and made the plant safer to operate because there is no continuous feed of chemicals, and handling concentrated sulfuric acid is avoided.

Even with this automatic cleaning system, some residue can build up in the stacks. It is recommended that the stacks be cleaned-in-place (CIP) on a routine basis, usually once a month. When the unit is switched into the CIP mode, the brine pump circulates a 5% muriatic acid solution through the membrane stack for scale control. If organics are present, a 5% salt solution (pH adjusted to 13)

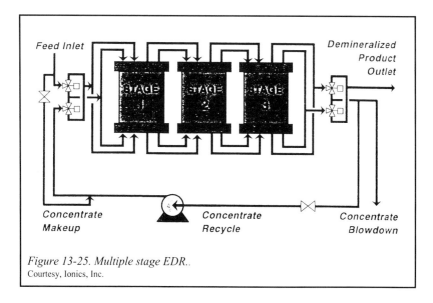

Figure 13-25. Multiple stage EDR..
Courtesy, Ionics, Inc.

will follow the acid cleaning. The complete cycle will take a maximum of 2 to 4 hours. A fail-safe system prevents the wash water, during the CIP operation, from contaminating the product water storage tank.

This unit is called the Aquamite system, and some idea of its performance is indicated in Table 13-8, giving the characteristics for the Aquamite XX (264,000-gpd output).

Following are some of the performance specifications supplied by Ionics, Inc. for its EDR system:

1. Continuous acid addition is not required. This, of course, eliminates the need for handling and storing concentrated acid.

2. As a result of no acid use, this system delivers a product that has a neutral pH. This can and does eliminate expensive items such as acid-resistant piping.

3. In most cases, sodium hexametaphosphate (SHMP) is not required as a sequestrant to eliminate calcium scaling since the polarity-reversing feature continuously keeps the membrane clean.❏

CHAPTER 14

ION EXCHANGE

Since the mechanics for the removal of sodium, hydrogen, and hydroxyl ions are basically the same, this discussion will first focus on the simplest of the three processes, sodium cation exchange (NaR).

Sodium Cation Exchange

This process is still referred to by some as a *zeolite softener.* Zeolite, a natural substance, is now seldom used, however, having given way with progress to man-made materials, resins (R), which are physically stronger and have greater capacities.

Figure 14-1 shows how these resin polymers can be visualized as tire chains on whose individual chain links are additions of spikes. These spikes can be likened to magnetic forces of equal strength. In cation resins, the spikes carry a negative charge to attract the positively charged cations; and in the anion resins, the spikes have a positive charge to attract the negatively charged anions.

Exchange materials with negatively charged skeletons are referred to as cation exchangers since they attract positive ions. The porous structure of ion-exchange materials permits water to permeate the particle, affording good contact with the exchange sites.

If calcium ions or cations of species other than sodium should approach a

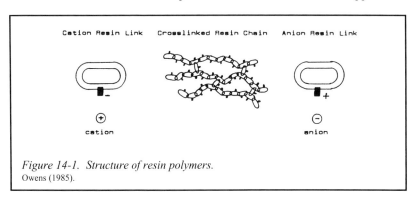

Figure 14-1. Structure of resin polymers.
Owens (1985).

cation-exchanger particle saturated with sodium ions, some of them may be captured by the exchanger. This would release sodium ions previously held at the exchange sites on the skeleton. An equilibrium is soon established between the distribution of these species on the exchanger and in the water surrounding the exchanger.

In essence, when sodium ions are regenerated with sodium chloride, they will become attached to the structure. In turn, these sodium ions will be replaced by the calcium and magnesium in the raw water, thus removing the hardness.

In the treatment of water, the basic concern is with six elements. Three of these — calcium (Ca^{2+}), magnesium (Mg^{2+}), and sodium (Na^+) — carry positive charges, and are referred to as *cations*. Three others — bicarbonate (HCO_3^-), sulfate (SO_4^{2-}), and chloride (Cl^-) — carry negative charges, and are referred to as *anions*.

Hardness in water is caused by the calcium and magnesium it contains; therefore, if some method for selectively removing the calcium and magnesium is available, soft water will result. Sodium cation exchange will meet this requirement.

Ion exchange can take place only with ions of the same charge. That is, positive ions will replace only positive ions, and negative ions replace only negative ions.

Note that this particular exchange material is constructed as depicted in Figure 14-1. Also observe that the positively charged sodium ion is attached to the negatively charged spike. This original sodium is furnished through a regeneration of the link with a concentrated solution (5% to 10%) of salt (sodium chloride).

As a result, the positively charged calcium and magnesium ions in the raw water will replace the sodium and will remain bound to the spike. The sodium ions will pass through the system in the now softened water.

If shown as a chemical reaction, the activity would appear as shown in the following equation.

$$Ca(HCO_3)_2 + Na_2R \longrightarrow CaR + 2NaHCO_3$$

The calcium is now attached to the exchange material (R), and it remains in the unit. At the same time, the sodium has become attached to the bicarbonate, and this sodium bicarbonate ($NaHCO_3$) remains in the treated water. The same reaction takes place with the $Mg(HCO_3)_2$ (magnesium bicarbonate), which is contained in the raw water, thus:

$$Mg(HCO_3)_2 + Na_2R \longrightarrow MgR + 2NaHCO_3$$

All of these calcium and magnesium salts will behave in a similar manner, and this is indicated in Figure 14-2.

When the capacity of the resin has been reached, it becomes necessary to regenerate the bed. In regeneration, the addition of ordinary salt, NaCl, causes the resin to release the calcium and magnesium ions it has captured, and this solution can then be flushed to waste. This reaction is shown in Figure 14-3.

In the regeneration and rinsing that follow, the calcium chloride ($CaCl_2$) and magnesium chloride ($MgCl_2$) are rinsed to waste and sodium ions again become attached to the skeleton:

$$CaR + 2NaCl \longrightarrow CaCl_2 + Na_2R$$
$$MgR + 2NaCl \longrightarrow MgCl_2 + Na_2R$$

Figure 14-4 illustrates a commercial softener. Note that the softener tanks are quite similar to a pressure sand filter. The sand bed of the filter is replaced with a bed of resin, and the necessary salt storage and brine measuring are added.

The operation of the softener is also quite similar to that of a sand filter. When the resin bed has become exhausted, the resin is backwashed, 5% to 10% brine solution is added, and the unit is then rinsed to waste until the rinse water hardness is zero.

The capability of resin to perform this exchange is limited. Each of the different resins is given a rating as to how much of this exchange can take place.

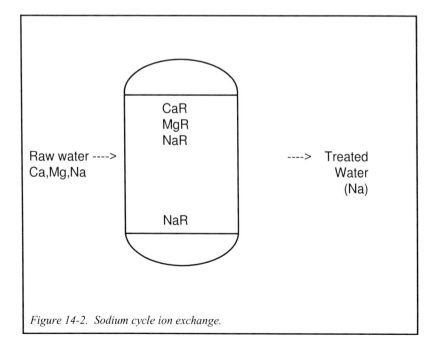

Figure 14-2. Sodium cycle ion exchange.

The rating capacity of natural zeolite ranges from about 2,500 grains per cubic foot (gr/ft^3) to some 6,000 gr/ft^3. The synthesized resins have capacities ranging from 6,000 to 32,000 gr/ft^3.

Design Parameters for Ion Exchange

The hypothetical water analysis below illustrates the process employed for designing an ion-exchange column. The calculations shown, not only for this particular resin but also for the hydrogen- and hydroxide-form resins that follow, are described only as a matter of academic interest. Any final design should be left to equipment manufacturers.

Chemical Substance	ppm as Shown	ppm as CaCO$_3$	gr/gal* as CaCO$_3$
Calcium (Ca)	60.0	150.0	8.8
Magnesium (Mg)	18.3	75.03	4.4
Sodium (Na)	13.8	29.95	1.8
Bicarbonates (HCO$_3$)	109.6	89.87	5.3
Sulfates (SO$_4$)	115.5	120.14	7.0
Chlorides (Cl)	31.9	44.98	2.6

*Since ion-exchange resin capacity is rated in gr/ft^3, use grains/gallon (gr/gal) in calculations: 1 gr/gal = 17.1 ppm.

a) Assume it is required to soften 25 gpm for 30 hours between regenerations,

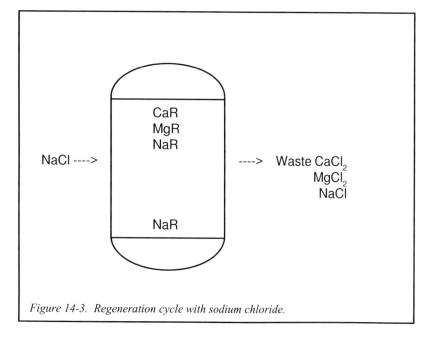

Figure 14-3. Regeneration cycle with sodium chloride.

then total volume of water between regenerations =
         25 gpm x 60 minutes x 30 hours = 45,000 gallons/regeneration

(Note: From this point on, some figures will be rounded off at the next highest whole number. Further, those sophisticated factors peculiar to different resins will be omitted. This is done so that there will be no distraction from the basic calculations. Actually the use of such factors does not affect sizing in most instances. The following can be considered as an average of the many different offerings made by different suppliers. See the resin rating assumption in section d.)

b) Sum of Ca + Mg = 13.2 gr/gal, round to 14 gr/gal

c) Total resin capacity required = 45,000 gal x 14 gr/gal = 630,000 grains

d) Resin rated at 30,000 gr/ft^3
         630,000 gr/30,000 gr/ft^3 = 21 ft^3 required

e) Assume flowrate of 5 gpm/ft^2 cross-sectional area. Then area of softener tank required = 25 gpm, at 5 gal/ft^2/min = 25/5 = 5 ft^2.

f) Diameter of softener tank = $\pi D^2/4$ = 5 ft^2 and,

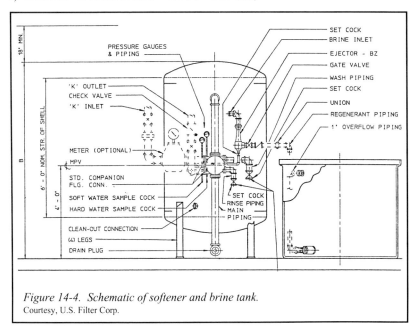

Figure 14-4. Schematic of softener and brine tank.
Courtesy, U.S. Filter Corp.

$$D^2 = 5/0.785 = 6.4 \text{ ft}^2$$
$$D = 2.5 \text{ ft} = 30 \text{ inches.}$$

g) Depth of resin bed = 21 ft³ / 5 ft² = 4.2 ft.

Since a resin bed will expand some 50% during backwash, the tank depth would then equal 4.2 ft + 2.1 ft, or 6.3 ft. To this should be added 18 inches or 1.5 ft, should a supporting bed be used. (See Figure 14-5.) Therefore, the length of the tank sidesheet should be 6.3 ft + 1.5 ft = 7.8 ft, rounded to 8 ft. Thus the softener tank dimension would be 8 ft sidesheet x 2.5 ft in diameter.

It might be well to consider using a tank with a 3-ft diameter. In such a case, the area of the tank would be 7 ft² and the depth of resin would then become 21 ft³/7 ft² = 3.0 ft. Thus the sidesheet dimension could equal 6 ft.

It might be wise to consider the use of duplicate units, and also the use of a resin with less capacity might be considered.

h) Depending upon the particular resin material, approximately 0.5 lb/1,000 gr of salt are required to regenerate this unit.

$$630,000 \text{ gr} / 1,000 \times 0.5 = 315 \text{ lb NaCl.}$$

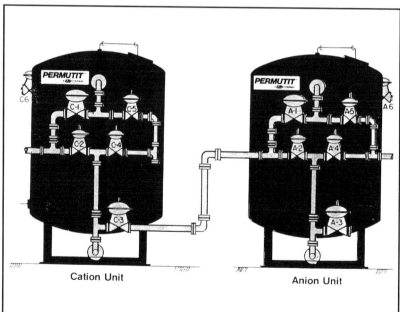

Figure 14-5. Cross section of ion-exchange unit.
Courtesy, U.S. Filter Corp.

i) Each gallon of saturated brine contains 2.5 lb of salt. Therefore 315 lb/2.5 lb/ gal = 126 gal of saturated brine required.

This brine is added to the resin as a 5% solution. Therefore, the saturated brine will be diluted to 315 lb salt/(315 lb salt + 8.33 gal) = 0.05

$$315 = 15.75 + 0.4165 \text{ gal}$$
$$299.25 = 0.4165 \text{ gal}$$
$$\text{gal} = 718.48$$

j) Whether a resin is being used for the first time or has been exhausted through its production of soft water, it must be regenerated before it is put back into service. This regeneration consists of backwashing, brining, and rinsing. First the resin bed is backwashed in the same manner as is a sand bed in a pressure filter. However, the flowrates are different and much lower for the resin (see Chapter 9). Also keep in mind that the backwash flowrate will vary with the temperature of the backwash water. The higher the temperature, the greater the flow required.

At any rate assume a backwash rate of 6 gpm/ft^2 of area, and in this case the flowrate would be 6 gpm/ft^2 x 5 ft^2 = 30 gpm.

Backwash continues for approximately 5 minutes or until it runs clear, and this will consume some 150 to 200 gallons of water.

Next, the brine addition consists of passing some 800 gallons of brine through the resin (from the top downward) for approximately 25 to 30 minutes.

Finally, the unit must be completely rinsed to waste, to remove all traces of the added brine. The rinse can be the raw water if it is not too dirty, or treated water if required. Rinsing at the softener rate would then require some 30 minutes at 25 gpm. The rinse must continue until the rinse water is of zero hardness.

Precautions

Turbidity or raw water should not exceed 5 ppm. Iron and manganese in the raw water supply should not exceed 1 to 2 ppm; otherwise, an iron-removal unit, described later in this chapter, should be considered.

Softened water tends to be corrosive, and a close watch should be maintained. If excessive corrosion is observed, it may be necessary to use a corrosion inhibitor downstream of the softener.

Remember that a softener removes only hardness. It does not remove alkalinity.

Hydrogen Ion Exchange

Hydrogen ion exchange is a method similar to that of the sodium cation exchange explained above.

If a resin adapted to hydrogen cation exchange (HR) is regenerated with a mineral acid such as sulfuric (H_2SO_4) or hydrochloric (HCl), the hydrogen (H) ion from the acid becomes attached to the resin. When water containing calcium, magnesium, and sodium ions comes into contact with the regenerated resin, the calcium, magnesium, and sodium ions will be replaced by the hydrogen ions. The following is a typical reaction:

$$Ca(HCO_3)_2 + H_2R \longrightarrow 2H_2CO_3 + CaR$$

Note that the calcium ion has been replaced by the hydrogen ion to deliver carbonic acid and CaR, and carbonic acid (H_2CO_3) breaks down into carbon dioxide (CO_2), which is a gas, and water:

$$H_2CO_3 \longrightarrow H_2O + CO_2$$

In a like manner the following reactions also take place:

$$Ca(HCO_3)_2 + H_2R \longrightarrow 2H_2CO_3 + CaR$$
$$CaSO_4 + H_2R \longrightarrow H_2SO_4 + CaR$$
$$CaCl_2 + H_2R \longrightarrow 2HCl + CaR$$
$$Mg(HCO_3) + H_2R \longrightarrow 2H_2CO_3 + MgR$$
$$MgSO_4 + H_2R \longrightarrow H_2SO_4 + MgR$$
$$MgCl_2 + H_2R \longrightarrow 2HCl + MgR$$
$$2NaHCO_3 + H_2R \longrightarrow 2H_2CO_3 + Na_2R$$
$$Na_2SO_4 + H_2R \longrightarrow H_2SO_4 + Na_2R$$
$$2NaCl + H_2R \longrightarrow 2HCl + Na_2R$$

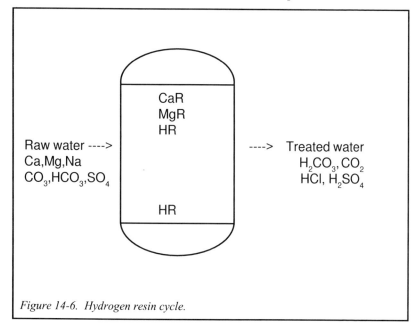

Figure 14-6. Hydrogen resin cycle.

This action is shown in Figure 14-6. Once the resin has become exhausted, it will require an acid regeneration, as in the example below and as shown in Figure 14-7.

$$CaR + H_2SO_4 \longrightarrow CaSO_4 + H_2R$$
$$MgR + H_2SO_4 \longrightarrow MgSO_4 + H_2R$$
$$Na_2R + H_2SO_4 \longrightarrow Na_2SO_4 + H_2R$$

The following description, which gives the salient information regarding the physical makeup of the exchange unit, is courtesy of Infilco Degremont, Inc.

General characteristics of an ion exchange unit: whatever the type of exchange, whether for softening, carbonate removal or deionization, each appliance normally consists of a vertical closed cylindrical container holding the resin. The latter can be placed in direct contact with the device collecting the treated liquid. The device may consist either of nozzles evenly distributed over a tray or of a system of perforated tubes of a suitable number and size. The resins may also be supported by a layer of inert granular materials (silex, anthracite or plastic beads), the layer being drained by the draw-off system.

Sufficient free space is left above the resin bed to allow it to expand normally (between 30 and 100% of the compressed volume depending on the type of resin) during back-flow expansion.

Both the water to be treated and the regenerant are admitted at the top of the

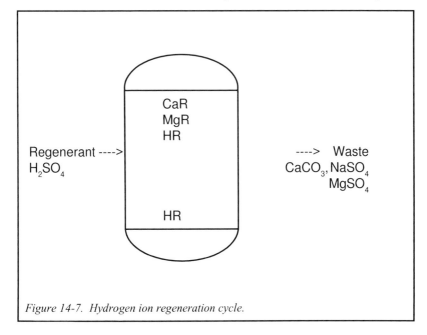

Figure 14-7. Hydrogen ion regeneration cycle.

container by a distribution system of varying complexity.

The appliance has an external set of valves and pipes for the various operations of fixing, expansion, regeneration and rinsing. The valves may be manually or automatically controlled, or can even be replaced by a central multiport valve. This equipment is identical with the resin softener in most respects. The only differences are as follows:

a) The regeneration is altered to feed acid instead of a brine solution.

b) Protective linings are required for all tankage, and rubber-lined or plastic piping must be used to transport the treated water to the point of use.

c) Arrangements for a somewhat more complicated rinsing of the unit must be arranged.

d) The effluent from the hydrogen resin exchange is of a very low pH, contains CO_2, and as a result is very aggressive.

e) The 66 Baumé sulfuric acid is extremely aggressive, and *extreme caution* should be exercised when it is used.

f) A cold shower should be located at the site of acid handling to shower the operator immediately should he be the victim of a spill or splash.

Rubber gloves should be used. Goggles should be used at all times when handling acid. Eye-wash equipment should be kept handy for rinsing out eyes.

A complete cycle of the exchangers consists of four basic operations:

● Treatment of water until resin becomes exhausted.

● At end of treatment run, the unit must be backwashed.

● Following backwash, the acid is applied from top downward and out the bottom to waste.

● All traces of the regenerant (acid) must be rinsed from the resin.

Design Parameters for a Cation-Exchange System

The following formulations required in the design of this unit are basically the same as those for the NaR unit described at the beginning of this chapter. (Again the reader is cautioned that consultation with the resin manufacturers on design parameters is a necessity.) The following units are used in the required calculations:

Gallons per minute - gpm
Gallons per minute per square foot - gpm/ft^2

Gallons per minute per cubic foot - gpm/ft^3
Grains per gallon - gr/gal
Kilograin - kgr
Kilograin per cubic foot - kgr/ft^3

a) Design flow is the amount of water to be treated per minute (gpm)
b) Gallons of water to be treated between regenerations =
design flow (gpm) x 60 x hours

c) Cubic feet of resin required =
$$\frac{\text{total cations (gr/gal) x gallons of water between regeneration}}{1,000 \text{ x rating of resin in kgr/ft}^3}$$

d) The quantity of resin having been established, it is now possible to determine the tank dimension:
Cross-sectional area of resin bed (ft^2) =
design flow (gpm)/permissible flowrate (gpm/ft^2)

e) Tank diameter^2 (ft) = area of resin bed (ft^2)/0.785
tank diameter = square root of above figure

f) Resin bed depth (ft) = resin bed (ft^3)/area of bed (ft^2)

g) Total depth of tank (sidesheet) (ft) =
Depth of resin bed (ft)
+ Depth of supporting bed (ft) if needed
+ Depth of freeboard (ft) - determined by the amount
of resin bed expansion (ft) resulting from backwash flow.

Thus if resin bed is 4 ft deep and bed expands 50% during backwash, 4 ft x 0.5 = 2 ft of freeboard above the normal bed depth.
Assume supporting bed, if one is used, is 2 ft. Then the total depth of supporting bed + resin bed + 2 ft caused by expansion = 8 ft. (If nozzle drain is used, no support bed would be required, and figure for supporting-bed depth can be ignored.)
There should be several feet above the expanded bed, so that in this case, the total length of sidesheet would be in the order of 10 ft.

h) Backwash flow = Backwash rate of flow (gpm/ft^2) x bed area (ft^2)

i) Acid application:
Lb of concentrated acid = lb acid/ft^3 x ft^3 resin

This acid is not applied at full strength, and must be diluted to meet resin requirements. The diluted acid is then applied at a specific flowrate to the resin bed.

j) The spent acid must be completely rinsed from the resin. To effect the best results, a so-called slow acid displacement rinse is followed by a fast rinse.

k) The total cycle time is approximately as follows:

Backwash	10 minutes
Acid application	45 minutes
Slow rinse	20 minutes
Fast rinse	40 minutes
Total	115 minutes

The following comments will add to the reader's understanding of this rather complex exercise:

- It requires approximately 10% to 15% of the volume of water treated to regenerate the HR unit.

- Backwash water can be raw water supply unless it has a high hardness. It is assumed here that the manufacturing supply is suitable.

- The treated water will contain carbon dioxide (CO_2) in proportion to the alkalinity of the raw water. The removal of this CO_2 will deliver a less aggressive water. As a result, a degasifier (Figure 14-8) should be included in the system if required.

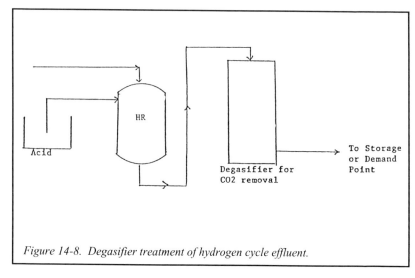

Figure 14-8. Degasifier treatment of hydrogen cycle effluent.

Hydroxyl-Form Anion Exchange

This unit is required to replace the anions in the effluent of the hydrogen cation exchanger with the hydroxyl (OH^-) ion. In this way, a totally demineralized water is obtained.

As the resin bed is becoming exhausted, the ROH is being replaced with RSO_4, RCl, and RCO_3, as shown in Figure 14-9.

$$H_2SO_4 + 2ROH \longrightarrow R_2SO_4 + 2HOH$$
$$HCL + ROH \longrightarrow RCl + HOH$$
$$H_2CO_3 + ROH \longrightarrow R_2CO_3 + HOH$$

When this resin is exhausted, it is regenerated with sodium hydroxide (NaOH) to replenish the OH ions, as shown in Figure 14-10.

The schematic in Figure 14-11 shows the equipment involved in total demineralization.

Design Parameters for a Hydroxyl Anion-Exchange Unit

Since the required cycles parallel those of the HR unit, the same procedures are followed here. There are, of course, differences and some are listed:

● The hydroxyl (OH) resin is an organic resin, different from the HR resin. As a result, different parameters and handling are required.

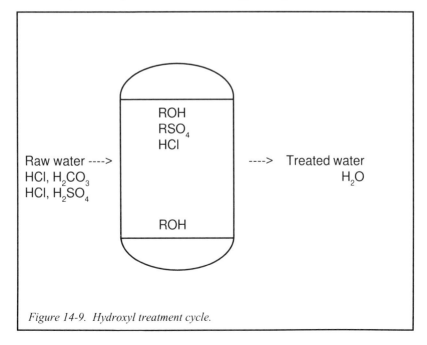

Raw water ---->
HCl, H_2CO_3
HCl, H_2SO_4

ROH
RSO_4
HCl

ROH

----> Treated water
H_2O

Figure 14-9. Hydroxyl treatment cycle.

● Caustic (NaOH) is the regenerant, as noted in Figures 14-10 and 14-11.

● The raw water usually found in a manufacturing plant can be used in all of the cycles of the HR unit. On the other hand, raw water is not used in the ROH unit. The ROH unit requires water that has been treated in the HR unit. As a consequence, once this volume of water has been determined, the volume of water treated in the HR unit must be increased to include the water required here. This means that the capacity of the HR unit will be increased by this amount. This usually means extra cubic feet of resin, which will increase the tank height.

The length of the cycles for a hydroxyl anion-exchange unit will vary from those of a hydrogen exchange unit. Following are approximate cycle times for a hydroxyl unit.

Backwash	10 minutes
Caustic application	70 minutes
Slow rinse	40 minutes
Fast rinse	40 minutes
Total time	160 minutes

Regeneration times:

Hydrogen exchange	115 minutes
Hydroxyl exchange	160 minutes

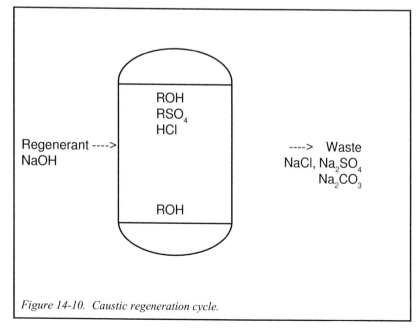

Figure 14-10. Caustic regeneration cycle.

In actual practice, the total time would not be 115 plus 160 minutes, since both units will be fast rinsed with the same flow. Also, expansion of the anion-exchange resin during backwash is considerably greater, thus a longer tank sidesheet will be required.

Mixed-Bed Resin System

In a mixed-bed system, as its name implies, the cation and anion resins are mixed as homogeneously as possible in a single vessel. This is in contrast to the two-bed cation-anion process previously described. The advantage of a mixed-bed system is that it will deliver a higher quality of effluent than that of HR plus ROH. This effluent is of a neutral pH, and the system usually provides the best silica and carbon dioxide removal. Figure 14-12 illustrates the physical aspects of the mixed-bed system.

The mixed-bed design is possible because the cation and anion resins have different densities. The cation resin, the denser of the two, requires a backwash flow of approximately 6 gpm/ft^2, while the backwash rate of the anion resin is usually less than 3 gpm/ft^2 of horizontal bed area. It is this difference in flowrates that makes it possible to separate the resins into two distinct layers.

At the end of this carefully controlled operation, the heavier cation resin will remain at the bottom of the vessel, while the anion layer will form a bed at the top of the cation bed. These beds are usually regenerated in place. Once this separation has occurred, the resins can be transferred if desired to separate

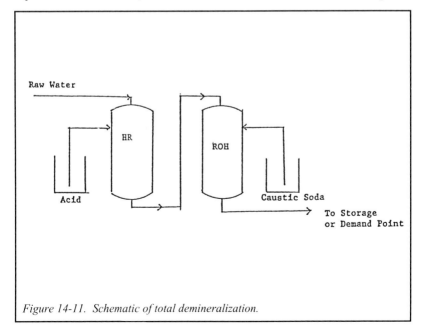

Figure 14-11. Schematic of total demineralization.

outside vessels for regeneration, as previously described in the cation process.

The transfer of resins is a controlled operation, but the rather simplistic depiction in Figure 14-12 indicates how this operation is carried out leaving the resins in the original vessel. Once the resins have been regenerated, they are thoroughly mixed again through air agitation furnished by the compressor. Close control must be maintained for such an operation to be successful.

Types and Combinations of Ion-Exchange Resins

Table 14-1 should bolster the warnings as to the parameters of design and operation of ion-exchange resin systems. The various designations and combinations shown differentiate among the resins as to factors such as performance, methods of regenerations, sodium and silica leakage, and raw water quality requirements.

Manganese Greensand Filters for Iron and Manganese Removal

Another process for the removal of iron and manganese is called *manganese greensand*. Although there are similar treatments available, most of the following information is obtained from literature issued by Hungerford & Terry, Inc., Clayton, N.J.

Ferrosand is manufactured to strict specifications to ensure the complete removal of iron and manganese using various oxidizing agents.

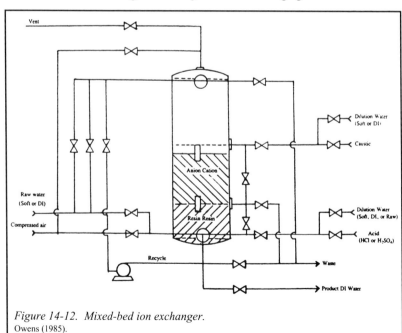

Figure 14-12. Mixed-bed ion exchanger.
Owens (1985).

Table 14-1
Ion-Exchange Resin Configurations

| | Inlet | | | Outlet | | | | Chemical Efficiency | |
| | Hardness | Alkalinity | Acidity | Water purity | | Silica | CO₂ | | |
	High	High	High	Moderate	High	Removal	Removal	Moderate	High
SAC/WBA			X	X		No	No	X	
SAC/FDA/WBA		X	X	X		No	Yes	X	
SAC/FDA/SBA		X		X		Yes	Yes	X	
WAC/SAC/FDA/SBA	X	X	X	X		Yes	Yes	X	
WAC/SAC/FDA/WBA/SBA		X		X		Yes	Yes		X
SAC/SBA/MB					X	Yes	Yes	X	
SAC/FDA/SBA/MB		X			X	Yes	Yes	X	
WAC/SAC/FDA/WBA /SAC/SBA	X	X	X		X	Yes	Yes		X
WAC/SAC/FDA/WBA/MB	X	X	X		X	Yes	Yes		X

WAC = Weakly acidic cation
SAC = Strongly acidic cation
WBA = Weakly basic anion
SBA = Strongly basic anion
FDA = Forced-draft aerator (or vacuum degasification)
MB = Mixed bed

The basic material is processed from what is commonly known as New Jersey greensand, but which is more correctly identified as the mineral *glauconite*. This is the same natural zeolite material originally used with the sodium replacement cycle for softening water.

Physically, the unit resembles to a great degree the zeolite softener mentioned above in that the system is comprised of a filter shell and a tank holding the regenerant solution. The unit uses potassium permanganate ($KMnO_4$) as a regenerant rather than the salt ($NaCl$) used in the softener, but the $KMnO_4$ is applied to the ferrosand in much the same manner.

Normally the tank contains only ferrosand, but if the raw water is such that high concentrations of iron and manganese are encountered, a dual-media anthracite/ferrosand (Figure 14-13) can be furnished.

The ferrosand itself is composed of black modular granules that have a density of about 5 lb/ft³. This material has a service flow rating that ranges from 2 to 5 gpm/ft², and requires a backwash rate of approximately 12 to 15 gpm/ft².

These parameters would equate to a unit that is 4 ft in diameter to treat 25 gpm. The backwash water in this case would be approximately 150 gpm. Note the similarity in this regard to the sand or anthrafilt filter.

Depending upon whether the raw water contains mostly only iron in combinations with manganese, or whether both ions are present in such amounts to require the removal of both ions, different methods of regeneration are recom-

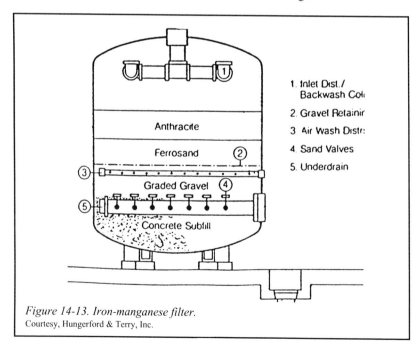

1. Inlet Dist./
   Backwash Col
2. Gravel Retainir
3. Air Wash Distr:
4. Sand Valves
5. Underdrain

Figure 14-13. Iron-manganese filter.
Courtesy, Hungerford & Terry, Inc.

mended.

The following chemical reactions are taken from the literature. Reactions involved in iron and manganese removal by potassium permanganate and manganese greensand include the oxidation of iron by chlorine (if used) and permanganate:

$$2Fe^{2+} + Cl_2 \longrightarrow 2Fe^3 + 2Cl$$
$$3Fe(HCO_3)_2 + KMnO_4 + 7H_2O \longrightarrow MnO_2 + 3Fe(OH)_3 + KHCO_3 + 5H_2CO_3$$

The oxidation of manganese by permanganate:
$$3Mn(HCO_3)_2 + 2KMnO_4 + 2H_2O \longrightarrow 5MnO_2 + 2KHCO_3 + 4H_2CO_3$$

And the reduction of any excess potassium permanganate by the manganese greensand (where Z represents manganese greensand [zeolite]) to manganese dioxide:
$$3Z^-MnO + 2KMnO_4 + H_2O \longrightarrow 3Z^-MnO_2 + 2KOH + 2MnO_2$$

Conversely, the oxidation of soluble iron or manganese by the manganese greensand when the oxidant demand on the raw water has not been fully met:
$$ZMnO_2 + Mn^{2+} \longrightarrow ZMn_2O_3 + Mn^{3+}$$

Continuous regeneration (CR) operation is recommended for well waters where iron removal is the main objective with or without the presence of manganese. This method involves the feeding of a predetermined amount of $KMnO_4$, usually in combination with Cl_2 directly to the raw water prior to the ferrosand filter.

Chlorine should be fed upstream of the $KMnO_4$ with a contact time of 10 to 20 seconds, if possible, using chlorine to produce the desired residual in the filter. The $KMnO_4$ should be fed to produce a "just pink" color filter inlet. This will maintain ferrosand in a continuously regenerated condition.

The concentrations of Cl_2 and $KMnO_4$ may be estimated as follows:
$$mg/L\ Cl_2 = mg/L\ Fe$$
$$mg/L\ KMnO_4 = (0.2 \times mg/L\ Fe) + (2 \times mg/L\ Mn)$$

Without Cl_2 and $KMnO_4$, demand may be estimated by:
$$mg/L\ KMnO_4 = (1 \times mg/L\ Fe) + (2 \times mg/L\ Mn)$$

As a matter of interest, this is the same material that is used with varying degrees of effectiveness for the removal of the obnoxious hydrogen sulfide found in some home well supplies.❏

CHAPTER 15

PRODUCT WATER

"There is no clear, simple quantitative way to describe the term 'water quality'. In practice, the quality of a water resource is determined by various measurements of physical, chemical and biological characteristics. Results of the measurements usually are compared with water-quality standards or criteria in order to judge the suitability of the water. There are several different sets of criteria, depending upon the intended use of the water; water that meets the criteria for one particular use will not necessarily meet the criteria for other uses." — NASQAN

There are a number of substances in municipal water that will be labeled as "impurities" if they are troublesome in product water. Quality control must determine whether or not clean, clear water furnished by the municipalities contains an excess of such substances. If so, it must be determined how best to remove enough of these deleterious substances so that a first-quality product water is obtained.

Water chemists have learned that regardless of the quality of the municipal supply, it is more than likely that further treatment will be required to meet specific parameters. These treatments could range from a simple zeolite softener for low-pressure boilers or cooling water; to a complex, sophisticated treatment required to obtain the high-purity water needed in the semiconductor (electronics) industry.

Each industry, be it paper, textile, chemical, pharmaceutical, or other, has promulgated its own standards. It is also interesting to note that even within some industries there are different requirements for different products. As an example of this, the Technical Association of Pulp and Paper Industry (TAPPI) standards are as shown in Table 15-1.

The purpose for this discussion has been to substantiate the statement, made in the opening paragraph of this chapter, that a municipality or private water purveyor cannot always and does not in every case deliver a water that meets the various individual requirements of industry. If the source water, which the municipality must treat before delivery to its customers, contains excessive amounts of some substances such as total dissolved solids, chlorides, or sulfates, for example, the final user will have to further treat the water for its particular needs.

Analyses Needed

The first order of business is to identify those substances that might be troublesome to product water. It will be learned, as previously stated, that usually the same substances are involved. The difference is that the amounts of these substances will vary from product water to product water. Therefore, the following list of substances to be analyzed for will be shown with no reference to their desired concentrations. A completed analysis will include some 20 or so characteristics and/or substances:

Suspended solids are those substances that are not in solution. They are solids that can be removed by filtration alone.

Turbidity could be classed as a subdivision of suspended solids. The difference between the two is that these solids are more finely divided.

Color is hue or tint imparted to an otherwise clear, clean water. The color is usually caused by soluble organic matter, iron and/or manganese, or some copper salts that have not been completely removed.

Iron (Fe). The natural source is from various minerals such as ferrous sulfide (FeS), ferric sulfide (Fe_2S_3), or iron pyrite (FeS_2); and also from certain rocks.

Manganese (Mn). The natural source of manganese is from certain soils and some minerals. This element is present in a dissolved state and, like iron, it will precipitate out as a dark brown, black, or rusty-looking solid.

Calcium (Ca) and magnesium (Mg). The original source for these is dolomite, which includes high-magnesium-bearing limestone, calcite, and gypsum.

Hardness (H), which is caused by compounds of calcium and magnesium, is a characteristic that is associated with soap usage — the harder the water, the greater the amount of soap required to form a lather.

Sodium (Na). Natural sodium is derived from feldspar, which is a crystalline mineral composed of complex aluminum silicates, sodium sulfate, and of course sodium chloride. The United States Geological Survey states that approximately 95% of one hundred of the largest U.S. municipal water supplies contains some 50 ppm or less of sodium.

Potassium (K). Generally, the natural sources are the same as for sodium. The United States Geological Survey states that 93% of one hundred largest U.S. municipal supplies contain 5 ppm or less of potassium.

Alkalinity: bicarbonates (HCO_3), carbonates (CO_3), and hydroxides (OH). Alkalinity as CO_3 is seldom sound in surface supplies, but it is not rare in well water, especially if the water has a high sodium content. Alkalinity as HCO_3^- is almost always present in surface, well, and municipal supplies.

Alkalinity enters water in the following manner: As rain and snow fall to the

Table 15-1
Pulp and Paper Standard

	Maximum concentration in process water				
	Fine Quality Papers	Kraft Papers Bleached	Unbleached	Mechanical Pulp Papers	Soda and Sulfite Pulp
Turbidity					
(in mg/L SiO_2)	10	40	100	50	25
Color					
(in platinum units)	5	25	100	30	5
Total hardness					
(French degrees)	10	10	20	20	10
Calcium hardness					
(French degrees)	5				5
Magnesium hardness					
(French degrees)					5
Methyl orange alkalinity					
(French degrees)	7.5	7.5	15	15	7.5
Iron					
(mg/L)	0.1	0.2	1.0	0.3	0.1
Manganese					
(mg/L)	0.05	0.1	0.5	0.1	0.05
Residual chlorine					
(mg/L)	2.0				
Soluble silica					
(mg/L)	20	50	100	50	20
TDS					
(mg/L)	200	300	500	500	250
Free carbon dioxide					
(mg/L)	10	10	10	10	10
Chlorides					
(mg/L)		200	200	75	75

See Tables 3-1 and 3-2 for conversion to other units.
Courtesy, Degremont *Handbook of Water Treatment*

earth, they will probably already contain some alkalinity, but in their descent they will absorb carbon dioxide from the atmosphere, and in doing so will form carbonic acid, a weak acid:

$$CO_2 + H_2O \longrightarrow H_2CO_3$$

As this water runs over the earth's surface, it will collect to form streams or lakes. Some will percolate through the surface to form wells. In areas over which this runoff occurs, the earth contains varying amounts of limestone, which is composed largely of calcium and magnesium oxides (CaO and MgO). The reaction between the carbon dioxide, the oxides, and water is as follows:

$$2CO_2 + CaO + H_2O \longrightarrow Ca(HCO_3)_2$$
and
$$2CO_2 + MgO + H_2O \longrightarrow Mg(HCO_3)_2$$

If sodium is present, then the following occurs:
$$2CO_2 + Na_2O + H_2O \longrightarrow 2NaHCO_3$$

The outstanding characteristic of alkalinity is its ability to neutralize acid.

Sulfates (SO_4). The natural source of sulfates is from the oxidation of various sulfide ores, and in some instances from industrial wastes. If sulfates are present in greater amounts than approximately 250 ppm and if they exist as compounds of magnesium or sodium, an off-taste can develop. It is fortunate that over 93% of the largest municipalities have sources of water that contain less than 100 ppm of sulfates.

Chlorides (Cl). The original source of chlorides is sedimentary and igneous rocks, and in some cases ocean water might contribute. As with sulfates, an off-taste ca develop if water contains more than 250 ppm of chloride existing as compounds of magnesium or sodium. Again, over 93% of municipalities have sources with less than 100 ppm of chlorides.

Fluoride (F). The original source is from fluorite ores, mica, and some ores that are found in granite and other hard rocks. The fluoride content in almost all large water supplies is 1 ppm or less. If the content is much greater, mottling of teeth will occur, but in the 1-ppm quantities, fluoride seems to be beneficial to the structure and resistance to decay of children's teeth.

Nitrate (NO_3). The primary sources of nitrate are legumes, plant debris, sewage, and nitrogenous fertilization. The occurrence of more than 5 ppm in public supplies is rare, and quantities greater than 45 ppm would require costly demineralization. Water containing more than this limit of nitrates, if used by senior citizens or babies (even unborn) makes them susceptible to the "blue-baby" syndrome, which impairs the circulatory system and can even be fatal.

Total dissolved solids (TDS). This is a measurement of all of the dissolved substances that exist in the water.

Free chlorine (Cl$_2$). By law, municipalities must add sufficient chlorine so that there will be a residual of 0.2 ppm at the tap. Any residual can impart taste; odor; and, if present in sufficient quantity, can cause a bleaching of a colored product.

Chloramine (NH$_x$Cl$_y$). The source of chloramines is the chlorination of ammonia-bearing compounds, whether these occur naturally or are added by the municipality as a means for controlling trihalomethane (THM). Ammonia is sometimes added to assure stability of the disinfecting property of chlorine in long distribution systems. In small amounts, chloramines will not affect taste and odor, but many municipalities deliver a supply in which taste and odor are noticeable.

Nonvolatile organic substances (NVOS). The sources of these are usually algae and pollution. Many are suspected carcinogens.

Volatile organic substances (VOS). The source of these substances is pollution; and these, when present, are found mostly in well supplies. Many are suspected carcinogens.

Taste and odor can be caused by hydrogen sulfide, by some volatile substances, or by other dissolved gases that result from the decomposition of organic material (algae). One of the most common causes is the formation of chloramines when a municipality adds ammonia to its supply in order to control trihalomethanes.

High-Purity Water

The quality of the product water required in manufacturing processes for the microelectronics, power generation, and pharmaceutical industries is no doubt equal to or superior to that required in any other industries.

As a matter of fact, the ultimate goal for quality control in these industries is to produce a product water that contains nothing but water. This water must be of such quality that it is described as *high purity*. By definition then it must contain a minimum or even an absence of all particles — be they soluble or insoluble; or be they chemical, physical, or bacteriological.

Water of this quality requires a parameter not mentioned previously. This unit of measurement, resistivity, is shown in a partial listing from a semiconductor manufacturer's standard:

Resistivity (25 °C - 77 °F)	17-18 megohm-cm
Particles (less than 0.1 μm in size)	2
SiO_2	3-5 ppb
Dissolved O_2	0
Total dissolved solids	Less than 1 ppm
Residue on evaporation	Less than 10 ppb

Note - The above values are used only for purposes of illustration. The reader is advised to check on up-to-date standards in any particular industry.

The only parameter that may require explanation is *resistivity*. Pure water is not a good conductor of electricity, and it is those substances (dissolved ions) that will increase the conductivity (and conversely, reduce the resistivity) of water. Conductivity will also increase directly with an increase in temperature. Consequently, the temperature standard selected is 25 °C (77 °F).

The unit of conductivity is the *mho,* and pure water will have a conductivity on the order of 0.00005 mhos. This value can also, of course, be written as 5 x 10^{-5} mhos, another rather awkward number to handle.

Because of the desire to use a more convenient unit of measurement, the reciprocal of the mho has been selected. This unit, called an *ohm,* is the unit of resistance.

$$1/mho = ohm \text{ and, } 1 \text{ micromho-cm(mho)} = 1 \text{ ohm-cm}$$

The processes selected to deliver high-purity product waters are varied and complex, and seem to incorporate combinations of all or most of those unit processes whose descriptions make up the content of this manual.

As a result, those concerned with the manufacture of these high-purity waters, be they quality control personnel, consultants, or operators, must have an intimate and thorough knowledge of these processes.

A reading of the literature shows any number of combinations of various equipments, but several catch the eye. A list of the unit processes incorporated in two such treatment plants appears as follows and in the order shown:

● Pretreatment
   a) Coagulation, filtration, or
   b) Activated carbon, or
   c) Multiple cartridge filter (5 μm)

● Cation exchange

● Vacuum degasifier

● Anion exchange

● Microfilter

- Ultraviolet
- Mixed-bed polisher demineralizer
- Reverse osmosis
- Ultrafilter

A second selection appears as follows:

- Activated carbon filter
- Cation exchange
- Degasifier
- Reverse osmosis
- Mixed-bed (primary) demineralizer
- Mixed-bed (polisher) demineralizer
- Ultrafiltration (0.22 μm)

Of course, also noted in the literature are those systems that incorporate only reverse osmosis or electrodialysis or ion exchange, followed in each case by ultrafiltration. Most of these finish with ultraviolet.

These latter systems are usually propounded by the various manufacturers. Some such systems incorporate pretreatment consisting of coagulation and filtration by activated carbon with microfiltration. Some manufacturers may suggest the use of only microfiltration as the sole means of pretreatment.

It follows then, that the quality of the raw water will have to enter the considerations, but it would seem that different designers have preferences of one type of process over another.

All of the above should convince everyone that the selection of equipment to produce high-purity water demands a careful consideration of all possibilities, and that experience will have a great bearing on the final selection.

As a result, the following sections will be thumbnail sketches of the "what and why" of several different product waters.

Microelectronics-Grade Water

Metals are good conductors of electricity; insulators are poor conductors of electricity. *Semi* means *half* or *partly*. Semiconductors are substances that conduct electricity better than insulators but not as well as metals.

It is this property that enables the arrangement of pieces of these substances to act like electron or vacuum tubes. These arrangements are called *transistors,* which are much smaller than the tubes. As a matter of fact, a large transistor is smaller than a dime, and the smallest transistors can be as small as the period at the end of this sentence.

The history of semiconductors began some 40 years ago with the exploration of the behavior of electrons in solids. Among a great number of solids examined, the most promising for semiconducting were silica and germanium. Since then the progress in the manufacture and use of semiconductors has been truly a modern phenomenon.

Silicon is the most widely used semiconductor. Silicon (Si), a nonmetal, is the second most abundant element next to oxygen. Silicon is found in nature combined with oxygen (O_2), this combination of silicon and oxygen is commonly known as sand (SiO_2).

Silicon as it is found in nature is not in the form in which it is used as a semiconductor, however. Silicon is obtained from the sand by chemical means. First, the sand is melted down. While it is in this molten state, a "seed" crystal of pure silicon is added, and as the mass cools down, the liquid silicon from the sand bonds to the crystal. When this coated crystal is pulled out of the melted mass a much larger crystal is formed. This extremely pure silicon is the basic material used as a semiconductor.

The next step in the process is to slice the mass into small, thin wafers that measure 4 to 8 inches across (10 to 20 cm), and varying thickness in the order of 1/100 inch (1/40 cm). Once the wafers are obtained, procedures known as *etching* and *doping* will convert them into the semiconductor *chip*.

These procedures are somewhat varied and, in some cases, trade secrets, but the following will give some idea as to how they are carried out.

Doping consists of the addition of certain chemicals (which now become impurities) to the pure silicon wafer. These impurities will become various parts of an electric circuit (such as transistors and switches).

The manufacture of semiconductor devices includes the etching of the wafers and the use of controlled thicknesses of surface oxides that alter the current-carrying or -insulating characteristics. These are carried out in a series of rather complicated maneuvers using light-sensitive coatings to transfer the circuitry pattern onto the water.

In each step of the manufacture, and before the device is encapsulated, there is a need for rinsing with water to remove any particulates and also to remove the last traces of the acid used for etching.

Consider, if you will, that the electrode patterns have a width of approximately 0.1 μm. If the rinse water contains any particulates or dissolved substances at all, this would, on evaporation of the water, leave such substances behind to alter the desired behavior of the device. It is for this reason that the highest quality of water is required.

Most manufacturers strive to obtain a water that contains less than 1 ppm of total solids, and that has a resistivity from 15 to 18 megohm-cm at 25 °C. This also assumes that the water has been sterilized for organic and bacterial control, and highly filtered. These specifications are for water that is probably as near to absolutely pure water as it is possible to attain; hence, the need for elaborate water-processing equipment.

In actual practice, the water quality will deteriorate on standing or when it is conveyed in pipes. Commonly, the main treatment plant will produce the quality of water required for the entire plant. This water then if required can be treated for the specific application in order to overcome any such deterioration. In such a situation, it is quite likely that disinfection will be required on a rather steady basis.

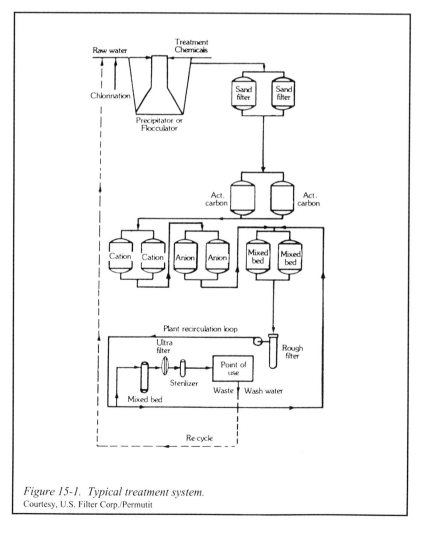

Figure 15-1. Typical treatment system.
Courtesy, U.S. Filter Corp./Permutit

Figure 15-1 is a typically basic treatment system. Note that almost all of the unit processes come into use. This particular system consists of the following steps in order:

1. Sterilization and/or coagulation
2. Filtration, dechlorination, activated carbon purification
3. Roughing demineralization (cation-anion exchangers or reverse osmosis)
4. Mixed-bed demineralization or polishing
5. Upgrading and maintenance of high-quality water in the ultimate water loop.

As a matter of interest to those who might believe that the use of high-quality water is novel to the manufacture of semiconductors, a similar water was required for the rinsing of the larger-caliber cartridge shells during World War II. At that time, however, processes such as reverse osmosis and electrodialysis were really not available on such a scale, and the rinse water was obtained from treatment through the cation-anion exchange process.

Water for Pharmaceutical Preparations
Unlike the consistently high-purity product water required in the manufacture of semiconductors, there seems to be a range of product water quality that can be adapted to the pharmaceutical industry. This does not detract from the desirability of high-purity water in all processes, but lesser than the high-purity water described previously can be and is used for laboratory use, rinse waters, clean-up water, and bottle washing.

The most critical demand is made on the water that is to be used for injection. This water must be free of pyrogens. In this sense, pyrogens are the excretions of bacteria in a solution (water). These excretions can cause fever in the patient receiving a medicinal injection carried in a water that is not pyrogen-free.

The quality of this water is specified in the U.S. Pharmacopoeia, and in order to guarantee the required sterility, this product water is usually distilled regardless of any pretreatment that precedes the distillation step.

It should become apparent, then, that the ideal water treatment process is that which can furnish product waters to suit the various water uses. For example, it is possible that the use of cation-anion exchange, reverse osmosis, or electrodialysis along with required pretreatments could result in an acceptable product water for clean-up water and other such uses.

At the same time, these treatments are excellent pretreatment for distillation. The use of such water will always result in more efficient distillation while eliminating or reducing the problems attendant to distillation.

The standards for these product waters are fixed by the requirement of the U.S. Pharmacopoeia, and they can be met by combinations of ion exchange, reverse osmosis, or electrodialysis. For some purposes it may be necessary to use dual systems of cation-anion exchange or reverse osmosis in series to achieve the required results.

Several suggestions are shown in Figure 15-2. Further, whichever treatments are used, they must produce product waters as specified by the U.S. Pharmacopoeia for each particular application.

If single-pass RO and/or single pass cation-anion exchange do not sufficiently treat the water, double-pass units or other combinations can be used.

The pretreatment might consist of ultrafiltration, coagulation plus sand filtration, activated carbon purifiers, or combinations of these processes, depending upon the quality of the raw water. It goes without saying that raw water supplies will require varying degrees of pretreatment for efficient operation of the main treatment processes.

(Not shown in Figure 15-2 is the use of the posttreatment equipment such as ultraviolet for point-of-use polishing.)

Boiler Feedwater
There is no doubt that one of the earliest treatment plants was used to furnish water for steam boilers. The problems facing the water chemist at that time continue to persist today. The one difference is that the problems are now more acute since the present boilers operate at much higher pressures and temperatures. Several of the persisting problems are identified below.

Scale formation. In boiler feedwater parlance, *scale* is defined as a deposit of salts that forms on the inside of a metal container when it is heated.

Everyone is probably familiar with the coating seen on the inside of a tea kettle

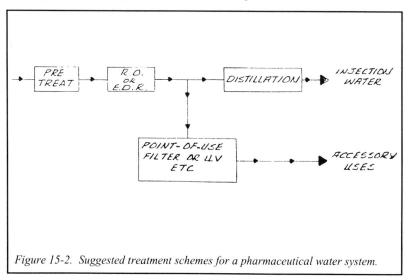

Figure 15-2. Suggested treatment schemes for a pharmaceutical water system.

after it has been in use over a period of time. Another example is the deposit that is formed inside a domestic hot-water heater.

Both of the above can be considered low-pressure boilers, if you will, and after the scaling becomes too severe these vessels are discarded. However, in the modern industrial boiler, some of which operate at pressures of up to and beyond 2,000 psig, the cost of replacement is out of the question.

Without going into the many ramifications of scaling, it is only necessary to say that it is caused by the dissolved soluble salts that are found in untreated water. The greater the amount of dissolved solids and the higher the operating pressure and temperature, the more severe will be the formation of scale. This scale can be so tenacious that it becomes extremely difficult, if not impossible, to remove. On the other hand, any insoluble solids carried by the feedwater result in the formation of a soft sludge that is easily blown down from the boiler.

The problems and damages attributable to scaling are many. Unless proper feedwater is used, the following will be two of the most troublesome:

● Lost efficiency of heat transfer, which results in loss of capacity and increased fuel cost.

● An unequal distribution of heat, which causes a resulting difference of expansion of the heating surfaces (tubes), with a consequential stress and distortion, and eventual failure of the metal.

The major cause of scaling is from the salts of calcium, magnesium, and silica (SiO_2); and as a result *external* treatment (treatment outside the boiler) is aimed to remove as much of these salts as possible.

Earlier treatments that are still in use are zeolite softening and cold or hot lime-soda ash treatment, followed usually by anthracite filters. Since such treatments do not remove all offending material, it is necessary to resort to *internal* treatment (treatment inside the boiler).

Internal treatment consists of the addition of various chemicals such as phosphate compounds and/or certain organic compounds (which are usually trade secrets). Such chemicals are used to inhibit both scaling and corrosion.

Of the salts mentioned, silica can cause serious problems not only in the boiler and boiler tubes proper, but also in the steam turbines used to generate electricity. Such silica can carry over with the steam to form silica deposits on the turbine blades, causing serious problems with the turbines.

Corrosion is the deterioration of metals by various chemical reactions, and both the rate and degree of degradation will increase with increased temperatures.

This phenomenon is caused by oxygen, carbon dioxide, and/or acids.

● The source of oxygen is usually found in the raw water supply to the plant, and it can be eliminated in part or wholly through deaeration.

● Carbon dioxide can also be found in the raw water supply, and this portion would be released during deaeration. However, the main source would be from the breakdown of bicarbonates in the supply water, as shown in the following equations:

$$Ca(HCO_3)_2 + heat\ (boiler) \longrightarrow CaCl_3 + CO_2 + H_2O$$
$$Mg(HCO_3)_2 + heat\ (boiler) \longrightarrow MgCO_3 + CO_2 + H_2O$$
$$2NaHCO_3 + heat\ (boiler) \longrightarrow Na_2CO_3 + CO_2 + H_2O$$

Of course the greater the alkalinity, the more carbon dioxide; and if these bicarbonates are not removed prior to the boilers, the carbon dioxide would be released inside the boiler where it would be most damaging.

The corrosive salts of calcium, magnesium, and sodium such as sulfates, chlorides, and nitrates will break down at the higher temperatures in the boiler to form sulfuric, hydrochloric, and nitric acids.

Just as in the case of the bicarbonates, these ions should be removed as completely as possible outside the boiler. If any of these remain in the feedwater after external cold or hot lime-soda ash treatment, their effect must be handled internally. This is performed by the maintaining of a predetermined alkalinity in the boiler water, such as through the use of caustic soda.

Caustic embrittlement is the breakdown or "cracking" of metal, usually at the riveted or welded portions of the boiler construction because it is at these points that the metal has been put under stress.

Other than the specifics of location, this problem has parallels to corrosion. Embrittlement is eliminated or minimized by proper treatment and by chemical dosing, either externally or internally.

The greater the operating pressure of the boiler, the greater the purity of makeup water that is required. (See Table 15-2.) To obtain this required purity for high-pressure boilers and nuclear generators, complete demineralization is the only safe procedure.

To effect the maximum removal of unwanted compounds and elements, it is necessary to resort to the high-purity treatment described at the start of this chapter.

Carbonated Beverages

Soft drinks, so labeled many years ago to differentiate them from spirits or "hard" liquor, inherently possess a number of excellent characteristics. These beverages, which are mostly water, are wholesome and pleasant to drink. They pass most readily into the circulatory system, and, in so doing, increase the volume of blood. They dilute food, which aids in the digestive process, and they aid in removing bodily wastes. Further, many now contain substantial volumes of fresh fruit juices, and from all indications this ingredient will be used in the future in even greater amounts.

In order to assure that soft drinks maintain these standards, all concerned must maintain the present high degree of water treatment. Also there must be methods and processes for handling newly-discovered or rather newly-legislated "impurities."

Although each of the major franchise beverage houses had promulgated its own set of standards before 1958, in this same year the International Society of Beverage Manufacturers issued the results (Helin, 1958) of a survey that was carried out in an attempt to "standardize the standards," so to speak.

The results of this study are shown in Table 15-3. Since some 60% of the parent franchisers and larger independent bottlers responded, it can probably be assumed that the average values would encompass most beverage water standards.

The chart is a summary of maximum, minimum, and average acceptable concentrations of the common constituents considered by beverage water chemists to be critical.

Table 15-2
Required Standards for Boiler Water

Working pressure	<15 bars	15-30 bars	30-45 bars	45-75 bars	75-100 bars
pH at 25 °C					
max	11.5-12	11.5-12	11-11.5	11	10.8
min	10.5-11	10.5-11	10.4-11	10.3-10.8	10.0-10.5
Total salinity, g/L					
max	2-4	2-3	1.5-2	0.5-1	0.1-0.5
NaOH, mg/L					
max	250-700	200-500	120-250	50-150	15-50
SiO_2, mg/L					
max	100-300	50-150	25-60	10-30	2-5
Na_3PO_4, mg/L					
min	50	50	50	20-40	10-20
max	—	—	120-150	50-150	20-50
Na_3PO_4/NaOH					
min	—	—	—	1	1

Note: The assumption is that the feedwater is always brought to zero hardness.
1 bar = 14.5 psig
Courtesy, Infilco Degremont, Inc.

Following is the conclusion reached by M.A. Helin, who wrote the report:
About the only standard that everyone agreed on was no taste or odor should be present or none that was objectionable.

To approach this subject from a scientific rather than a statistical basis, the scope of such an undertaking would be tremendous when all variable elements that may adversely affect the many different beverages are taken into consideration. For example, no two syrups are the same just as no two waters are alike. It is conceivable that temperature, pH or any one of the multitude of dissolved substances both organic and inorganic nature and their concentrations may affect one beverage to a greater or lesser degree than another. It appears, therefore, that to establish single maximum quantitative standards, each beverage would have to be studied separately.

A perusal of this report will draw attention to several other conclusions:

● The only references to organic matter content are the values of 0.4 ppm (average) and 0.0 ppm (median). There was really no attempt to obtain any other meaningful information. As a matter of fact, the value of 0.4 ppm

Table 15-3
Beverage Water Survey
(all amounts in ppm)

	Max	Min	Avg*	Median Avg*
Turbidity	10	0	2.3	2.0
Color	20	0	4.8	3.5
Organic matter	5	0	0.4	0
Taste & odor	0	0	0	0
Chlorine	0.2	0	0.03	0
Alkalinity	130	0	70	50
Sulfates	900	0	240	225
Chlorides	525	0	210	225
Iron & manganese	1.8	0	0.4	0.1
Copper	0.05	0		
Calcium	500	25	182	150
Magnesium	650	0	160	80
Sodium	900	500		

*The *arithmetic average* of a series of numbers is the sum total of the individual values divided by the number of values.
Example: Average of 2, 4, 6, 8, 12 = (2 + 4 + 6 + 8 + 12)/5 = 32/5 = 6.4
*The *median average* is the number that divides the series of numbers into two equal groups of items.
Example: In the series of 2, 4, 6, 8, 12, the value 6 separates these values into two groups: There are 2 and 4 in the one group and 8 and 12 in the other group.
Helin (1958); Courtesy, International Society of Beverage Manufacturers.

(average) was probably a value that indicates "oxygen consumed." If this be the case, it is not really a true measure of bacterial quantity. Such a value could include other oxidizable substances such as iron and manganese.

● One set of standards could not be established for the various constituents considered objectionable in beverage water.

● This survey indicates that the beverage water must at least meet the standards as set forth by the United States Public Health Service (USPHS) as they existed at that time except for organics and alkalinity, perhaps; and as has been previously stated, the USPHS and EPA standards are the same.

Although it is known that the differences in product waters from different beverages at the present time are very small, the caveat is issued that *quality control and operators must strictly adhere to the standards as set forth by their parent companies.*

Fortunately, chemically pure water is not essential. Even under controlled laboratory conditions, chemically pure water (i.e., water that contains nothing other than two atoms of hydrogen chemically combined with one atom of oxygen) is difficult and costly to produce. Waters suitable for human consumption are not chemically pure, and need not be so. However, drinking waters often contain substances in concentrations that, although not detrimental to human health, are detrimental to the taste, appearance, or shelf life of soft drinks.

As a result of further study, the parameters shown in Table 15-4 were more or less tacitly accepted as a general guideline for beverage water, and they remain

Table 15-4
General Beverage Water Standards

Turbidity	0 ppm
Color	0 ppm
Taste	0 ppm
Odor	0 ppm
Chlorine	0 ppm
Chloramine	0 ppm
Organic matter	0 ppm
Alkalinity	50 - 75 ppm
Iron & manganese	0.1 ppm
Sulfates	250 ppm
Chlorides	250 ppm
TDS	500 ppm
THM	75 ppb

so to this day.

Although it has previously been stated that chemically pure water is not required for a first-class beverage, studies are now being conducted by some bottlers to determine the efficacy of the use of such water. In some products, reverse osmosis is now actually in use for the reduction of sodium in water to be used for "low-sodium" drinks. It is also being used for reduction of organics such as THM, and other substances.

Other experimentation using high-purity water for other purposes is being studied privately by some soft drink franchises. In the meantime, the basic treatment described under "Conventional Treatment" (Chapter 9) is used. In addition, ultraviolet is becoming popular in the soft-drink industry for bacterial control.❏

CHAPTER 16

CONTROLS AND INSTRUMENTS

Controls and instruments are required for water treatment systems because these systems generally need to be cycled every few hours or days. Ion-exchange demineralizers and softeners need regeneration several times each day. When a cycle occurs, a number of valves and controls need to be operated in a carefully planned sequence in order to direct the flows in the right direction at the right time.

Regardless of the process, one of the first decisions to be made is whether the operation is to be manual, semiautomatic, or fully automatic. If the various stages of treatment are to be operated manually, manpower will be required to start and stop the complete plant, open and close valves, start and stop pumps, and operate chemical feeders. With a manual system, a lapse on the part of the operator can prove to be expensive.

In an automatic system, the entire plant and all of its functions will be stopped and/or started by signals that indicate activity in a piece of equipment, such as a change in water level in a tank or loss of head across a filter bed. Further, if any of the parameters such as final alkalinity, conductivity, chlorine residual, or effluent turbidity are not as required, such signals can be used to stop and start a plant, or to alarm the operator that attention is needed.

The semiautomatic system operates as does the automatic system except that at least one of its functions is performed manually. In both of the automatic modes, the sequential positionings of filter and carbon tower valves and regeneration equipment are automatic, by means of solenoid valves located in the main control panel.

To avoid error and to reduce manpower, plant designers and suppliers choose the design most appropriate to the application.

To most simply introduce this subject, the discussion will be process-oriented, beginning with a study of the conventional treatment system described in Chapter 9.

Conventional Treatment Plant
Figure 16-1 illustrates a conventional treatment plant. Each of the unit

processes and the functions within those unit processes will be discussed individually.

Chemical Reactor. The functions that require control for the operation of the reactor are handled by several units. A raw water controller is used to maintain a constant flowrate for the raw water coming into the reactor. This flow is equal to the rated capacity of the reactor.

The controller is a combination pressure regulator and shutoff valve. A constant flow is maintained through the use of a downstream orifice that is sized to deliver the desired flow at the controller pressure. A simple control that either opens the valve wide or closes it tightly is a two-way valve. This type of valve is preferred because it gives a nonmodulating control — it cannot pause in a partially open position. Once the control is turned to either position, operating fluid flows into or out of the cover chamber (see Figure 16-2) until the valve is open or closed. Ordinary three-way valves usually are not satisfactory because they require too much force to operate. Preferred methods of assisting valve operation include solenoids, pressure-activated diaphragms, and level floats. In one particular case, a float switch activates a solenoid valve that applies pressure to and removes it from the diaphragm.

Chemical feeders. Coagulant, chlorine, and lime feeders generally operate at

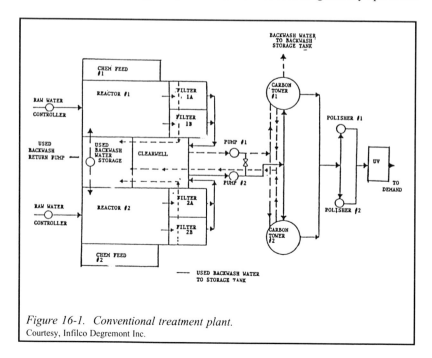

Figure 16-1. Conventional treatment plant.
Courtesy, Infilco Degremont Inc.

a constant rate, dependent on the flow and composition of the raw water. Chapter 17 illustrates a case where the quantity of lime is regulated through the use of a conductivity measuring device.

Sludge recycle and sludge discharge. The volume and concentration of the recirculated sludge are dependent upon the quantity of sludge formed by the chemical reactions and by the quantity of sludge discharged to waste. These values are determined empirically during the early stages of operation, and a close control is required for best reactor effluent.

Rotor-mixer. This equipment at times simulates a low-head, high-capacity pump. Manual adjustments are made for the particular water and its treatment one time during the early stages of operation. The only other control required is that of stopping and starting this motor. These functions are actuated by a series of float switches located in the clearwell, which is the storage tank for the filter effluent.

Clearwell. The rotor-mixer described above is actuated by the level of water in the clearwell. Figure 16-3 shows the primary activation points. Note that there are three float switches. These devices will actuate valves that can be pneumatically, hydraulically, or electrically driven. They will also actuate motors.

When the tank is empty, the mid-level switch closes an electric circuit that will open the inlet water valve and start the chemical feeders. At the same time, the reactor rotor drive, sludge recirculation, and sludge valves to waste are activated. This operation will continue until the water level reaches the high-level switch, at which time all of the above functions are shut down.

As water is drawn from the clearwell, its level will drop until it reaches the

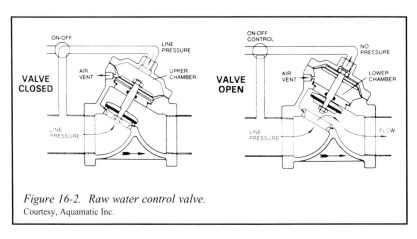

Figure 16-2. Raw water control valve.
Courtesy, Aquamatic Inc.

mid-level switch, at which time all of the aforementioned functions start up again. This cycle will keep repeating because the capacity of the treatment plant is greater than the production demand.

The low-level switch shown in the sketch will come into play only if the water drops to this level. In such a case, it will shut down the booster pump that delivers water to the carbon towers, and thence to production. Should this pump be allowed to run dry, cavitation with its most undesirable results will occur.

Other than for equipment malfunctions, the only interruptions that might occur in this cycle are when the filters and the carbon towers are backwashed.

Gravity filter. Water flows from the reactor to these filters by gravity. This filtration is required to assure that no unsettled particles that are present in the reactor effluent will reach the carbon particles. As the filtration progresses, these solids form a mat on and just below the surface of the filter media and, in so doing, restrict the flow through the filter bed. Since the flow of water to the filter is applied at a constant rate, the filter must operate at this same rate. If it does not, the level of water will rise in the filter cell and, should this continue long enough, water will eventually overflow to waste.

Rate-of-flow controller. To counteract such overflow operation, a modulating valve is placed in the filter outlet line. This valve will open gradually as the water level rises to allow more water to pass through the filter; thus, a fairly constant water level is maintained through the filter cell.

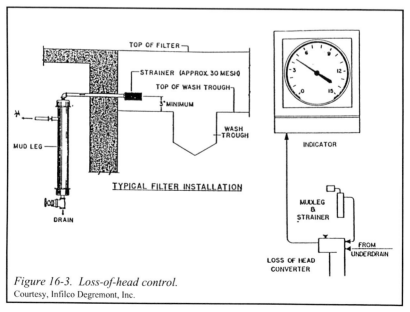

Figure 16-3. Loss-of-head control.
Courtesy, Infilco Degremont, Inc.

The reverse is also true when the filter is clean. A clean filter will allow too much water to pass through, with the danger of leaving all or at least part of the filter bed exposed and, subsequently, will allow air-binding.

In this case, as the water level in the cell drops, this valve will throttle to maintain the proper level. This modulating valve is called a *filter rate-of-flow controller*. One of the earliest types of this modulating valve used a direct mechanical linkage between a float riding the level of the water in the filter cell and a butterfly-type valve located in the filter effluent line. However, as the flows increased, so did the size of the effluent valve until it was not really possible to effectively operate in this manner.

The controls in use today more or less follow the basic pattern shown in Figure 16-4. Note that a venturi tube (or orifice) is part of the system. A venturi tube is a device with a constricted passage (throat) that raises the velocity and lowers the pressure of the water that passes through it.

In such a system, the differential pressure that is created between the entrance to the tube and its throat (constricted section) is converted to a pneumatic energy that will modulate any changes in the water level in the filter cell. The meter shown can be of the indicating and/or recording type, or the controller can operate without either.

Loss-of-head gauge. During the operation of the filter, there exists a difference between the pressure at the water level in the filter cell and the pressure at the bottom of the filter bed. The pressures at the two points are indicated by the loss-of-head gauge shown in Figure 16-5.

There is a small difference in pressure when the filter is clean, and this difference increases as the filter run progresses. As a result, a loss-of-head

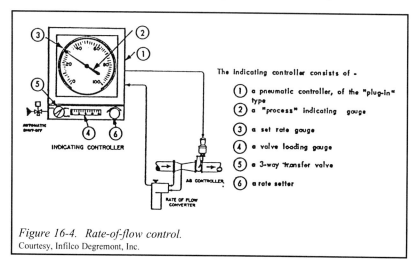

The indicating controller consists of -

① a pneumatic controller, of the "plug-in" type
② a "process" indicating gauge
③ a set rate gauge
④ a valve loading gauge
⑤ a 3-way transfer valve
⑥ a rate setter

Figure 16-4. Rate-of-flow control.
Courtesy, Infilco Degremont, Inc.

gauge can be used to indicate the time and necessity for washing the filters.

The differential is measured by the converter, which will sound an alarm or issue an output signal for any metering. This same signal can be used to initiate backwash, and also to open and close the effluent controller, which in this instance would result in a tight shutoff.

Activated carbon tower. The filtered, heavily chlorinated water that is stored in the clearwell is dumped through the carbon towers to production. From an operating standpoint, these are the principle differences between the gravity filters and the carbon towers:

● The gravity filter cell is an open vessel in which the operating level must be controlled.

● The carbon tower is a sealed vessel that can withstand whatever pressure is required to deliver the effluent to the points of demand. As a result, there is no need to resort to a level controller. Rather, the flow is controlled solely through the use of a rate-of-flow controller located in the purifier outlet line.

● Solenoid valves that are activated through the control panel automatically position the valves on the tower.

Figure 16-6 shows one type of valve used on the carbon tower. When pressure is not applied to the upper chamber, the diaphragm is off the valve seat

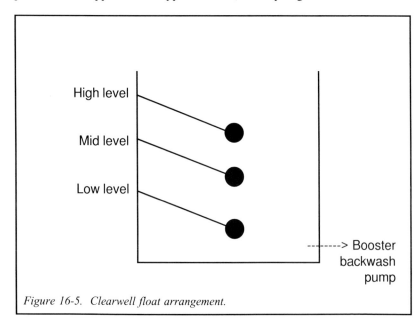

Figure 16-5. Clearwell float arrangement.

(open). When pressure is applied, the diaphragm is positioned onto the valve seat (closed) to shut off the flow. This pneumatic pressure is turned on and off through a pilot located in the control panel and through a solenoid valve (Figure 16-7).

A solenoid is a coil of wire that carries an electric current. When this coil is energized, it behaves like a magnet to attract the lever that positions the small valves in the solenoid valve assembly. This device will alternatively apply or divert the pressure to open or close the diaphragm valve.

An alternate choice to the diaphragm valve is the butterfly-type valve, which can be activated by the same pilot and solenoid valve mentioned above.

Backwash and rinse functions. A filter backwash can be programmed to be activated automatically by one of several signals but, as a matter of convenience, the operator may prefer to activate the backwash manually. When a filter is being backwashed, the reactor that supplies it is automatically taken out of service.

This backwash water that comes from the clearwell is pumped to the bottom of the filter, and the flow is upward through the filter bed into the filter washthrough. From this point, the flow is by gravity to the used-backwash storage tank.

The quantity of this flow is controlled by the design of the backwash pump or by means of an additional flow controller. The duration of the backwash is

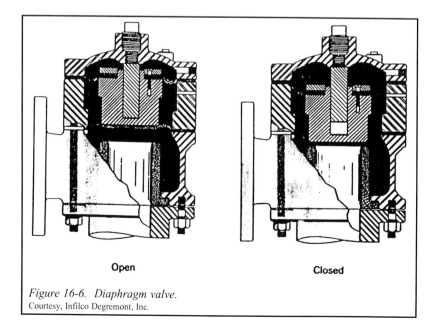

Open   **Closed**

Figure 16-6. Diaphragm valve.
Courtesy, Infilco Degremont, Inc.

controlled through the use of a timer located in the control panel.

The same procedure discussed above is followed for backwash of the carbon tower, except that this dirty backwash water is delivered under pressure to the used-backwash water storage tank. At the end of the backwash cycle, the tower is returned to operating mode.

The filter rinse-to-waste function can be programmed automatically, but in most cases this practice is being abandoned because of new filter designs.

The carbon tower rinse is in the same category, except that it would be wise to include a rinse in the design to assure that the tower effluent to the demand points is free from any chlorine residual. (A chlorine monitor is used to alarm the operator or to keep this unit out of the line.)

Used-backwash storage tank. This tank is similar in design to that of the clearwell except that it contains only two float switches. It should be large enough to contain the backwash water resulting from at least two of the filter cells (1A and 1B in Figure 16-1), and preferably large enough to contain also the backwash water for cells 2A and 2B (Figure 16-1). When this tank is full, one of the float switches will have risen to its upper level, and in so doing will activate to shut down the reactor inlet line, chemical feeder, and sludge accessories. At the same time, it will actuate the used-backwash return pump to send this dirty water to the reactor.

No chemical feed is required since only the suspended solids must be removed from this previously chemically treated water. As this tank empties, the lower float switch will actuate to shut down the backwash return pump, and the treatment plant reverts to normal operations. When a filter is backwashed, the reactor that supplies this filter must be taken out of service. As a result, the treatment plant at this stage will be producing only one-half of the production demand. Further, since the backwash water is filtered water, the treatment plant

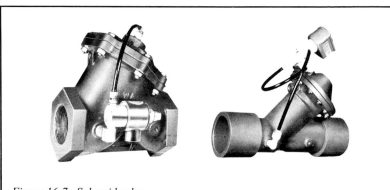

Figure 16-7. Solenoid valve.
Courtesy, Aquamatic Inc.

is actually supplying less than one-half the production demand. Although a reactor is not removed from service when a carbon tower requires backwash, a similar supply problem may exist.

To avoid these situations, it is most important that the design of the clearwell is such that sufficient filtered water is stored to meet both the backwash and production demands.

Several other possibilities come to mind. For one, the use of a third filter and a third carbon tower might be considered. Under this scenario, these units would be used to replace the filter or tower taken out of service for backwash, or they can "float" on the line.

A second clearwell for storage of carbon tower effluent could be included. However, such would result in the possibility of recontamination, and would definitely require an additional pumping.

Pumps. The return backwash pump is sized so that it does not exceed the capacity of a single reactor (see Chapter 17). The two booster backwash pumps are identical. During operation, one pump is used to boost the filtered water through the carbon towers and polishing filters to the points of production demand.

The second pump is used for backwashing purposes, and once a week or so, the pumps are alternated so that over the course of time each of these pumps will have been in service an equal period of time.

This illustrates, in part, what functions are required to operate semiautomatically or automatically. Further, it illustrates the interaction of timers located in a control panel; float switches located in the clearwell and used-backwash water storage tanks; and the use of pneumatically operated valves that are activated by solenoid valves.

This particular equipment with this program of achieving these results (as shown in Figure 16-1 and described above) is only one way of doing so. Different manufacturers of such control equipment offer different possibilities.

The details of the control panel (programmable controller) are highly specialized matters, proprietary information that manufacturers of this electronic equipment are somewhat reluctant to release. Some discussion of these controls follows, however.

Ion-Exchange Controls

At the risk of oversimplification, it might be said that except for the materials of construction, the design of the equipment for these ion-exchange processes is quite similar to that of a pressure sand filter and of the activated carbon tower (described in the last section and in Chapter 9).

The essential operational difference between the two is that for ion exchange it is necessary to regenerate each of the resins with its appropriate chemical.

This is accomplished through the use of chemical storage tanks and eductors and/ or pumps.

All such units can be operated in the semiautomatic or automatic modes through float switches, solenoid valves, the use of pneumatically or hydraulically operated valves, or the interaction of timers located in a control panel.

For ion exchange the control panel is designed to sequentially position the valves in the front piping, as shown in Figures 16-8 (a-d). First, note that only two valves are open during each step of the regeneration, which occurs in the following order:

a) Backwash

b) Applications of regeneration chemicals

c) Rinse

d) Service

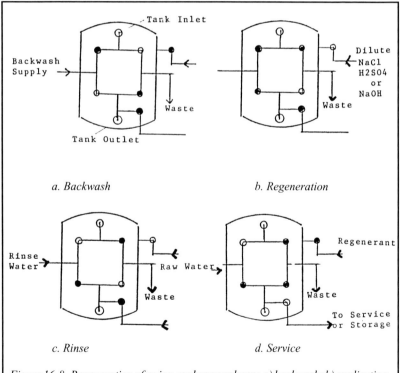

Figure 16-8. Regeneration of an ion-exchange column: a) backwash; b) application of regeneration chemicals; c) rinse; d) service. (White circle = open valve; filled circle = closed valve).

The complete assembly for the valves shown in Figure 16-8 includes the following parts:

a) Valves (diaphragm or butterfly)

b) Valve actuator

c) Three-way solenoid valve, which can be located either at the diaphragm or butterfly valve or in the panel. The solenoid valve is actuated from the control panel, and it directs air pressure (if pneumatically operated) to open or close the valve.

Rate-of-flow controllers or pressure-regulating valves and orifices are used to control backwash, rinse, and service flows. All of these features are common to all ion exchangers. The description that follows of a brine valve and brine tank for a sodium cycle (water softener) can be considered as one idea of regenerant addition for all ion-exchange units. (See Figure 16-9.) Of course, the regenerants will be different for other exchangers.

1. With the brine tank full and the brine control valve in the float-up position, the valve is ready to open when vacuum is applied.

2. When a softener starts its brine operation, water passes through the ejector, creating a vacuum in the brine line to the brine control valve.

3. With the control in the float-up position and a vacuum present, the brine control

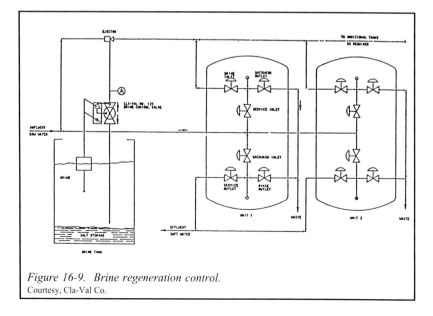

Figure 16-9. Brine regeneration control.
Courtesy, Cla-Val Co.

valve opens to allow brine to flow through the ejector into the softener tank. Brine flows through the softener and out to waste.

4. As brine is withdrawn from the brine tank, the float ball travels down the float rod. Upon contacting the lower float stop, the pilot control is positioned to vent the cover of the brine control valve to atmosphere. Atmospheric pressure on the valve diaphragm closes the valve against the vacuum and stops the flow of brine to the softener.

5. When the flow of brine stops, only fresh water continues to flow through the ejector for brine displacement or slow rinse.

6. After completion of the brine and rinse operation, the line to the brine control valve is under pressure, causing the valve to open. Water then flows through the brine control valve, refilling the brine tank.

7. As the brine tank fills with water, the float ball travels up the float rod. Upon contacting the upper float stop, the pilot control is positioned to apply pressure to the cover, closing the brine control valve.

The service cycle for the sodium cycle softener is terminated when the effluent hardness begins to increase. The service cycle on the other cation and/ or anion exchangers will be determined by one of the following:

● High conductivity in effluent

● High silica in effluent

● High sodium in effluent

● Variation in pH in effluent

● Total gallons treated

Each of these parameters can be measured either by rate of flow or by chemical characteristics, and such measurement can be used in automatically starting or stopping the process. The length of backwash and rinsing are controlled by timers located in the control panel.

The above discussion has not included features such as storage tanks for treated water or for undiluted regenerants but has merely highlighted features that might be included in plant design of the demineralizers.

Programmable Controllers: Concepts

Most of the foregoing is really not much more than a review of several methods of treatment process control. A thorough study of this matter will reveal that the theory of process control encompasses much more.

Before any further discussion is possible, it is necessary to become familiar with certain concepts that are peculiar to control systems:

Loop. A loop can be defined as one basic unit or function of a process control system. Using the conventional treatment system as an example, it contains a number of loops. This number will depend upon what the designer expected to accomplish. Following are some possibilities:

a) Reactor and accessories loop

b) Gravity filter loop

c) Activated carbon tower loop

d) Clearwell loop

e) Backwash and storage loop

f) Booster / backwash pump loop

Other functions that are combinations of the above can also be considered loops.

If the gravity filters are analyzed, it will be noted that controls are required for the following:

a) Opening and closing of valves for filter, backwash, and rinse to waste

b) Loss of head

c) Level control

d) Rate of flow

Each of the filters will require control and each of these controllers is referred to as a *local* control. In the final scheme, these are supervised by the *master* control, sometimes called the *remote* control. These have been discussed in some detail.

To familiarize the reader with yet another concept, the rate of flow is analyzed:

Set point. The set point is a reference to the level at which a system, function, or unit it is desired to operate. If it is desired to operate filters at 80% of design flow, for instance, this value is set on the appropriate gauge. This is called the set point. This set point and the actual flow through the filter are sent to the flow controller. The controller compares these two values, and if the actual flow through the filter is, say, 70% of design flow, the controller will open wider to increase the flow. If the flow through the filter is greater than the set point, the controller will throttle to decrease the flow to arrive at the set point value.

The power used to open and close a rate-of-flow controller is compressed air or water pressure. When all such loops that control the various plant functions are coordinated and designed into a single master controller, a programmable controller will exist.

Varieties of Programmable Controllers

A programmable controller covers a wide range of possibilities, extending from electromechanical units via static logic control devices, to systems employing minicomputers. The programmable automatic unit is one such system.

The two basic types of controllers are electromechanical and solid state. Solid-state systems offer more reliability and the ability to monitor. Electromechanical systems consist of switches, adjustable timers, relays, and other accessories required to carry out the desired operations of the units. These can be carried out (as previously discussed) either semiautomatically after manual initiation, or after a signal transmitted by a flowmeter or other measuring device.

A programmable system replaces switches and relays with a solid-state programmer. It can monitor the influent qualities of the incoming water and thus direct the quality of the effluent water. The programmable controller can be designed so that the operation of the treatment system is completely automatic. However, in all of the systems discussed, some phases of some of the cycles (loops) are usually initiated manually.

The types of control will vary, depending upon the applications and/or operation. As a result, a number of different types of control are available, and a few of them are listed below.

Proportional control is one that could be adapted to maintaining the level of water in a filter cell. This could replace the system discussed in the conventional treatment plan. The accuracy of this method depends upon the range (band) of control desired.
Integral mode is similar to proportional control except that range of the proportion is not too critical.

Feedback control is one in which there is a continuous checking of the effluent. This impulse is then used to adjust feed in order to obtain desired results.

For instance, when feeding sulfur dioxide to neutralize chlorine or chloramine, it is essential that there be no sulfur dioxide remaining in the treated water. The feed of sulfur dioxide is controlled by the chlorine residual in the water after the sulfur dioxide has been added.

Feed forward control is one that does not use feedback control. This practice is quite common, for example, in the proportioning of alum or other chemicals to a varying flow of water.

The above are only some examples of the many types of control systems that are available. Each is certainly worthy of consideration for qualifying applications.❏

CHAPTER 17

ACCESSORY EQUIPMENT

Centrifugal Pump

A centrifugal pump is the most commonly encountered pump in a water treatment installation. This pump operates on the principle that if circular motion is imparted to the water, the water will tend to move away from the center of the applied force. This is called *centrifugal energy,* the simplest example of which is seen when a weight attached to a string is swung around. As the weight swings, the string becomes taut because of the tendency of the weight to pull away and fly outwardly. The force that causes the weight to pull away is called *centrifugal force.* The amount of force developed will increase with increased length of string and/or increased speed of rotation.

A simple centrifugal pump consists of arms to spin the water and power to spin the arms. The arms are called *vanes,* and a series of vanes is called the *impeller* (Figure 17-1). The power is furnished by an electric motor. The pump consists of a motor that drives a shaft to which is attached the impeller. To contain the water so that it does not fly off in all directions, the impeller is housed in a *casing.*

So that the shaft and consequently the impeller rotate smoothly, the shaft is supported and kept in alignment with bearings (see Figure 17-2). To keep water from leaking through the gap where the shaft enters the casing, a mechanical seal or stuffing box is provided. A stuffing box is nothing more than a box stuffed with packing that forms a waterproof gasket to seal the gap.

Pump maintenance required of the operator is proper lubrication, proper packing, and maintenance of a clean impeller. Careful thought will need to have been given to the selection, placement, and piping of the pump. The operator is advised to study the manufacturer's manuals and to follow closely any recommendation made on use and maintenance.

Pump Lubrication. Some newer models of motor pumps or reducers use bearings that require no lubrication. For all others, however, bearings must be lubricated according to the manufacturer's recommendation. Ball bearings are quite sensitive to under-lubrication, which will eventually cause wear and the need for bearing replacement. For most worm gear units such as those located in the reaction tank mechanism, a shot of grease should be applied every 100 to 150 hours of running time.

Over-lubrication can lead to excessive heat in the bearings and consequently reduced bearing life. Bearings will show some temperature increase following lubrication, but the temperature should return to normal after about 8 hours. If higher-than-normal temperatures persist, chances are that too much lubrication was applied. In that case it will become necessary to remove grease. (See manufacturer's instructions.)

Stuffing boxes are generally packed with a braided asbestos packing that has been impregnated with graphite. However, the same effective seal can be achieved through the use of a shaft seal constructed of a synthetic rubber. Under all circumstances consult manufacturer's recommendation for types of asbestos packing or sealing rings. Either the asbestos packing or the shaft seal is held in place against the casing by means of the packing gland.

When starting a new pump installation or following a repacking of the pump, the packing gland should be loosened until the packing gland nuts are only finger-tight. This will allow the box to leak rather heavily for some 10 to 15 minutes. This leakage should be reduced gradually by tightening the nuts a few turns at a time. Continue to tighten the nuts evenly, a little at a time, until eventually the leakage from the stuffing box is about 20 to 30 drops per minute.

Under no circumstances should the packing gland tightened to the point that there is no leakage of lubricant onto the shaft. This leakage is the element that coats, lubricates, and seals the shaft so as to prevent air leakage into the pump.

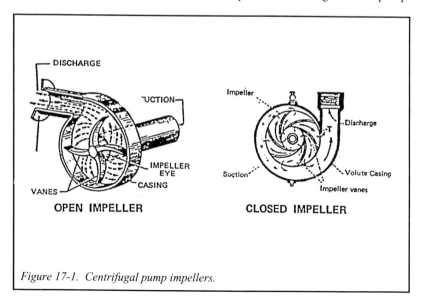

Figure 17-1. Centrifugal pump impellers.

For these installations, the sealing water is fed to the stuffing box through an interior passage from the pump casing.

Impeller maintenance. There will be times when, because of heavy coatings of calcium carbonate and/or magnesium hydroxide and magnesium carbonate on the impeller, the capacity of the pump might be reduced. This condition, which takes a few years to develop, can result from lime treatment of the water. If all other means to achieve correct pump pressure and capacity fail, it is a safe bet that the coating on the impeller must be removed. This can be done with the use of a mild inhibited acid.

Pump Selection

The pumps used in a conventional treatment plant (and also, perhaps, in other processes) are used both for boosting the water through filters and carbon towers, and also for backwashing these same units. Consequently, a pump must be selected to serve these dual requirements. The graph of pump curves in Figure 17-3 demonstrates the sizing of a pump.

Start at the blip, the heavily-shaded inverted L at the center of the graph. This corresponds to the desired maximum output: 500 gpm with a head of 100 ft. Note that the intersection of these lines falls between 20 and 25 horsepower; therefore a 25-hp motor will be required to drive the pump.

The heavily shaded curve just above the blip determines that at 3,500 rotations per minute (rpm), the impeller diameter is 6 1/2 inches. This value is found by following this curve to the inner scale at the extreme left.

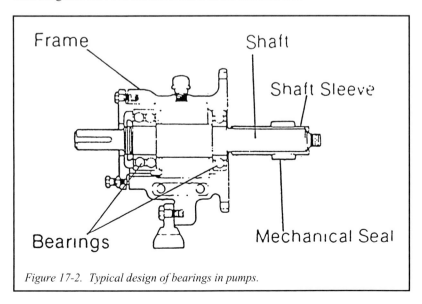

Figure 17-2. Typical design of bearings in pumps.

Since 300 gpm is the desired average operating flow, the point at which the impeller curve intersects the 300-gpm capacity line determines that the discharge pressure at this flow will be about 150 feet (70 lb/inches^2).

The parabolic curves indicate that the pump efficiency is ±63% at 300 gpm and approximately 58% at 500 gpm. These efficiencies are below the usual 70% or so, but keep in mind that the pump was selected to serve two purposes.

Another parameter that can be determined from this graph is net positive suction head required (NPSH$_R$). These are shown in Figure 17-3 by the broken vertical lines, which range in values from 6 ft to 30 ft. At 300 gpm, which is the operating flow in the system under consideration, this value is about 10 ft. Likewise, at the 500-gpm flow, which is the required backwash water flow, the net positive suction head available (NPSH$_A$) is 14 1/2 ft.

The NPSH$_R$ can also be calculated (Figure 17-4). The available pressure (head) is equal to the sum of atmospheric pressure and the height of the water over the centerline of the pump:

$$14.3 \text{ lb/in}^2 + 2.3 \text{ ft H}_2\text{O/lb pressure} = 33.9 \text{ ft} + 8.0 \text{ ft} = 41.9 \text{ feet}$$

From this value one must remove the head required by the vapor pressure (vp) and the head loss caused by the friction of the maximum flow (500 gal through a 4-inch gate valve and 4-inch dome):

$$\text{vp at } 40°\text{F} = 0.25 \text{ ft}$$
$$\text{vp at } 50°\text{F} = 0.4 \text{ ft}$$
$$\text{vp at } 60°\text{F} = 0.6 \text{ ft}$$

The head loss caused by valves and piping
= 4.2 ft, therefore NPSH$_A$ = 33.9 ft + 8 ft - (0.6 + 4.2) = 37.1 ft.

These calculations show that even should the water level in the tank drop to 1 foot above the centerline of the pump, there would still be sufficient NPSH$_A$ to avoid cavitation. Cavitation will not only reduce the pumping efficiency but, even worse, cause possible damage to the pump.

As an illustration of pump selection, assume that it is desired to select a pump that will deliver 200 gpm through a filter and purifier to a point 100 ft away at which a pressure of 50 lb/inch^2 is required.

It is first necessary to select a pipe diameter. At 200 gpm, a 4-inch pipe will result in an acceptable flow velocity of 5.05 ft/sec. At 200 gpm the loss caused by friction will be 4.29 ft/100 ft of pipe. From Figure 17-5 note that approximately 150' of pipe is required.

Because there will differences in the front piping, it will be assumed that the pressure loss through the front piping will be 20 ft through each tank. Further, it should be noted that the pump must lift the water to an elevation of 10 ft to get the water to the tanks.

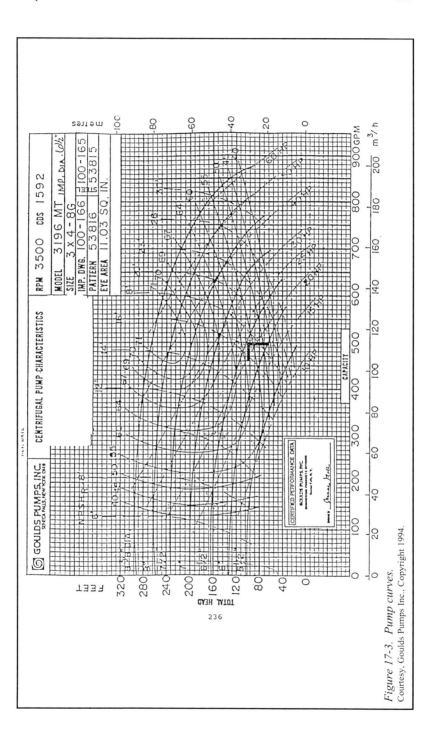

Figure 17-3. Pump curves.
Courtesy, Goulds Pumps Inc., Copyright 1994.

Adding the above will give the pressure required to push the water through the tanks to the demand point 100 ft away.

$$150 \text{ ft of 4-inch pipe} = 150 \times 5.05/100 = 7.6 \text{ ft}$$
$$\text{Loss through tanks} = 2 \times 20 = 40 \text{ ft}$$
$$\text{Static head} = 10 \text{ ft}$$
$$50 \text{ lb/in.}^2 = 50 \text{lb} \times 2.3 \text{ ft/lb} = 115 \text{ ft}$$
$$\text{Total} = 172.6 \text{ ft}$$

Now refer to the previously discussed pump curve (Figure 17-3). Starting at 200 gpm on the lower horizontal scale, go vertically to 172.6 on the vertical axis, and note that the lines will intersect somewhere between the 15- and 20-hp curves. In such a case the 20-hp motor is used.

A check of this selection can be made through the following calculations:

$$\text{Horsepower} = (\text{gpm} \times 8.33 \text{ lb/gal} \times \text{head in ft}) \div 33,000 \text{ lb/min}$$
$$\text{Horsepower} = (200 \times 8.33 \times 172.6) \div 33,000 = 8.7 \text{ hp}$$

The 9-hp pump is at 100% efficiency. The pump from the previous selection exercise delivered at about 60%, which is a little low, but this entire exercise has been an estimate.

Vacuum Gas Chlorinator

Another commonly encountered accessory in a typical water treatment system is a vacuum chlorinator (Figure 17-6). The heart of the vacuum chlorinator is a device called an *injector*, which consists of a main pipe and an auxiliary pipe. As water flows through the main pipe, a vacuum or suction is created in the auxiliary pipe, and this will draw the chlorine gas from its storage cylinder. As

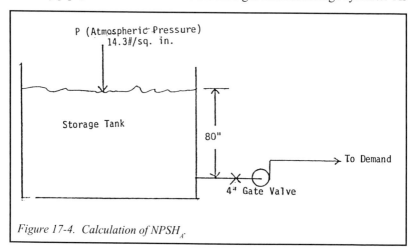

Figure 17-4. Calculation of NPSH$_A$.

it does so, it also mixes the chlorine gas and the water to form a chlorine solution. The amount of vacuum created will depend upon the quantity of water through the main pipe.

An automobile carburetor is an injector. In a carburetor, the flow of air in the main pipe creates a suction that pulls in the gasoline from the auxiliary pipe (fuel line) and forms a combustible mixture of air and gasoline. When a carburetor is "adjusted," it is to control the air-fuel mixture to give the best combustion in the cylinder chamber.

In a similar manner, it is necessary to build into the gas chlorinator certain control devices so that the strength of the final chlorine solution is as desired. It is also required that certain safety precautions be included since chlorine is a dangerous chemical.

The gas chlorinator used in product manufacturing plants is capable of feeding some 100 lb of chlorine gas per 24 hours. It is of the manual, variable vacuum control type.

Main pipe chlorinator controls. In the main pipe furnishing the water supply to the chlorinator, the following will be found:

● Shutoff valve to take the system off-line, followed by pressure regulator

● Strainer to remove any fine particles, which tend to clog any small orifices in the system

● Check valve, usually built into the injector control valve, and required to avoid backup of any chlorine solution into the water supply line
This portion of the chlorinator supply is illustrated in Figure 17-7.

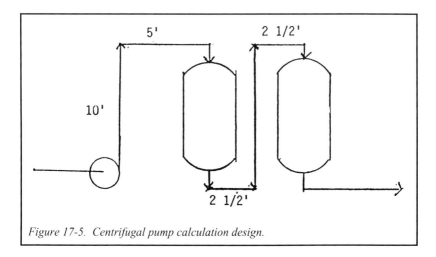

Figure 17-5. Centrifugal pump calculation design.

Chlorine gas line controls. The following controls are built into the auxiliary line:

● Chlorine gas shutoff valve, to isolate the chlorine gas cylinder from which the chlorine gas is obtained.

● Gas pressure regulating, to maintain desired gas pressure to the system.

● V-notch variable orifice, to control the amount of chlorine gas feed to the injector. On these particular units, to change the amount of gas fed to the system, a manual adjustment must be made.

● Vacuum regulating valve to control the amount of vacuum in a similar manner as does a pressure-regulating valve in a water line. This portion of the system can be portrayed as shown in Figure 17-8.

Interconnecting the lines from Figure 17-7 and 17-8 gives the flow scheme for the entire gas chlorinator system, as depicted in Figure 17-9. In addition to the features shown in Figures 17-6 through 17-9, chlorinators can also include a pressure-vacuum relief valve and a check valve, which on this unit is built into the injector unit.

The pressure-vacuum relief valve serves a dual purpose. Should the chlorine supply be exhausted or shut off, or should the vacuum valve fail to function properly, the vacuum on the upstream side of the V-notch orifice will increase. At a preset vacuum, the diaphragm in this valve will open to relieve the excess vacuum. If for some reason, the chlorine pressure regulating valve does not properly seat, a slight pressure will develop in the system, and the diaphragm in this valve will be forced in the opposite direction to relieve this pressure through a vent to the outside atmosphere.

The check valve in the combination injector / check valve is installed in order

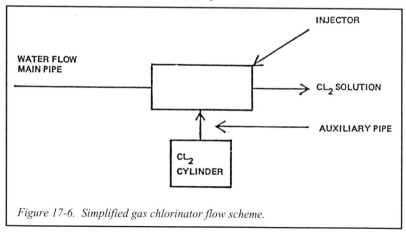

Figure 17-6. Simplified gas chlorinator flow scheme.

to avoid any backflow into the water supply vent.

Aeration, Deaeration, and Degasification

Aeration is a process in which air and water are brought into intimate contact with each other. As simple as this may sound, there always seems to be some confusion to those not familiar with this subject matter. For example, the same piece of equipment is called at various times an *aerator,* a *deaerator,* or a *degasifier* (sometimes even a *decarbonator*). At the same time the process is then called aeration, deaeration, or degasification. This would indicate that the same piece of equipment is defined and called by the result its use is expected to produce.

Forced-draft aeration. Figure 17-10 shows a simple type of aerator. Water enters the unit at the top, and as it cascades over slatted trays on its path to the bottom, it is met by an upward flow of air, either forced up as in Figure 17-11,

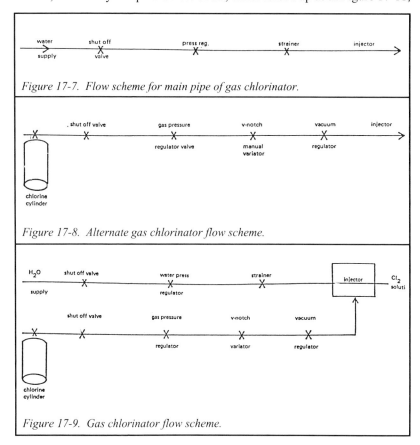

Figure 17-7. *Flow scheme for main pipe of gas chlorinator.*

Figure 17-8. *Alternate gas chlorinator flow scheme.*

Figure 17-9. *Gas chlorinator flow scheme.*

or sucked up from the top. In either case the contact time between air and water would remain the same. The only real difference would be the placement of a blower (at the bottom) or an induced-draft fan (at the top).

To give some general idea of size, a unit capable of handling 500 gpm would be approximately 7 ft square by approximately 12 ft high.

Normally some 6 ft³ of air per gallon of water will be required so that about 3,000 ft³ of air would be furnished by a 1-hp blower. Such an aerator might be of wood (fir or redwood) construction, which results in good economy. The trays can be removed for cleaning.

Other types include the cascade-type deaerator and the vacuum deaerator. The latter is probably the only unit that can completely remove entrained air or other gases. It is a simple operation that requires little maintenance, is highly efficient, but is relatively expensive.

Conductivity Control for Lime Treatment
The variation of alkalinity in the water supply to a product manufacturing plant usually does not exceed ± 5%, because the municipality that furnishes the water

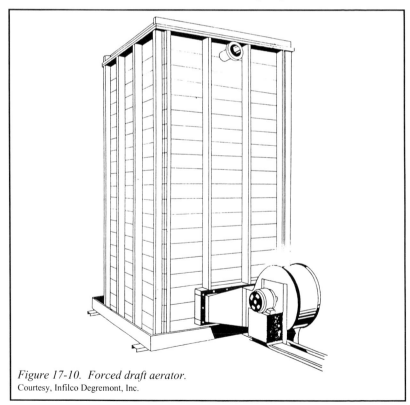

Figure 17-10. Forced draft aerator.
Courtesy, Infilco Degremont, Inc.

is able to control the treatment to its raw water in a consistent fashion. Any large variations in chemical composition of the supply are usually seasonal. In such cases, the plant has learned from past experience or has been forewarned by the municipality. Because of these factors, the operator will normally be able to control the alkalinity of the water used in the process without too much difficulty.

There are, however, cases in which the above does not hold true, and there are instances in which a supply (private or municipal) can vary widely in chemical characteristics. Consider, for example, the variation as shown in Table 17-1 of four different sources that make up the supply for a particular large water district.

There are plants that do not use municipal supplies, but for one reason or another must depend on their own private well supplies. In these cases, the supply may come from one or more wells located in the same well field. Although situated in the same field, these wells will in all likelihood deliver waters with different chemical compositions.

In any case, it may become necessary for either a municipality or a plant to blend supplies from several wells in order to achieve some measure of consistency. This is difficult if not impossible, and the desired result is usually not obtained. Without some type of automatic control for the feeding of the lime used for alkalinity control, any attempt to deliver a desired alkalinity for the beverage water (to use an example) would be almost impossible.

Table 17-1
Well Water Comparisons

Substance	Units	Source A	Source B	Source C	Source D
Silica	mg/L SiO_2	7.4	7.2	13.6	8.3
Calcium	mg/L Ca	80	18	48	58
Magnesium	mg/L Mg	31.5	10	18.5	24
Sodium	mg/L Na	108	37	46	84
Potassium	mg/L K	4.7	2.0	3.2	4.1
Carbonate	mg/L CO_3	4	0	0	0
Bicarbonate	mg/L HCO_3	144	73	124	135
Sulfate	mg/L SO_4^{2-}	297	43	116	191
Chloride	mg/L Cl	96	47	49	86
Nitrate	mg/L NO_3^-	<0.1	0.8	1.9	0.2
Fluoride	mg/L F	0.34	0.09	0.37	0.27
TDS	mg/L	701	202	359	524
Total hardness	mg/L $CaCO_3$	329	86	196	244
Total alkalinity	mg/L $CaCO_3$	124	60	102	111
Carbon dioxide	mg/L CO_2	1.2	1.0	2.9	3.1
pH		8.31	8.10	7.86	7.86

This is really not as far-fetched as might be imagined. The curves in Figure 17-12 show hourly variations in the alkalinity of a supply delivered to one plant. They also show the alkalinity of the treated, filtered water.

In order to treat raw water with varying alkalinities, Infilco Degremont has developed a proprietary process that combines a sludge blanket type of unit with a control based on the difference in conductivity between the raw water and this same raw water to which lime has been added for alkalinity reduction.

A ratio controller for lime feed as used with the Densator unit for alkalinity

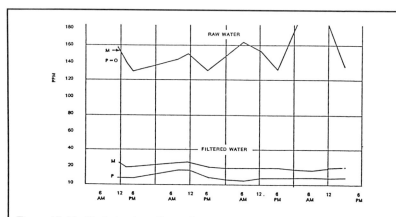

Figure 17-11. Variation in well supplies.
Courtesy, Infilco Degremont, Inc.

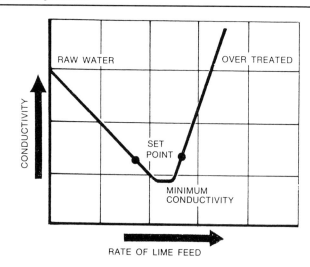

Figure 17-12. Relationship of conductivity to rate of lime feed.
Courtesy, Infilco Degremont, Inc.

reduction consists of two conductivity-sensing cells and a ratio controller (Solubridge). One cell is placed in the stream of raw water. The second cell is placed in the primary reaction zone. As the raw water contains the alkalinity-causing bicarbonates, the addition of lime will form insoluble calcium carbonate, which precipitates. This causes a reduction in dissolved solids and a corresponding reduction in conductivity (see Figure 17-11). It is in this manner that a ratio between treated and raw waters conductivities is established, and the ratio controller will call for more or less lime to maintain the set ratio required to deliver the desired alkalinity.

Note from the curve in Figure 17-12 that the process is not controlled by removal of the bicarbonates. Rather, the portion of the raw water that flows into the primary zone and is overtreated with soluble hydroxides (from lime) is what supplies the conductivity differential between it and the untreated raw water. This excessive alkalinity is, of course, neutralized by the portion of the raw water that flows into the secondary zone of the reactor.❑

CHAPTER 18

WASTE TREATMENT

As more environmental restrictions are being imposed upon the manufacturer, the choice of a water treatment process becomes of greater importance. This choice may well depend not only on the results desired, but on the type, amount, and cost of handling the waste resulting from the particular process. Further, all of these considerations have to be made while keeping in mind the public's increasing concern about the environment.

Fortunately, there is progress in both the technology and equipment required so that there are more varied options, but it must be kept in mind that what may be a good solution in one location may well be a detriment in another.

To this end, a thorough knowledge of various water processing techniques is essential so that the best selection can be made. The wastes from these treatment processes, added to the waste flows detailed in this chapter, will determine the required waste treatment process.

For the purpose of this discussion, assume that there is no connection from the plant to a municipal sewer. Rather, assume that all of the plant wastes are poured into a beautiful small lake or pond or flowing stream located near the plant. Do you have any idea how long it would be before this beautiful body of water would be turned into a real nuisance? Rest assured it wouldn't take long.

Actually, such a disposal measure is neither practical nor permissible, and the day is here that municipalities do or will soon start to charge not only for the amount of discharge to their sewers, but also for the amount of pollutants that are contained in this discharge.

In the case of the pond assumed earlier, lack of oxygen would cause decay and decomposition of the organic substances (sugars) contained in the plant discharge. This form of degradation is labeled *anaerobic* (no oxygen).

Self-Purification
On the other hand, if the same plant waste were to be discharged into a rapidly flowing stream of sufficient size, this stream, although also polluted, would eventually return to its original state (assuming that no further pollution is added in the meantime).

This phenomenon is called *self-purification*. So long as sufficient oxygen has been naturally absorbed by the flowing stream, this purification process is labeled *aerobic* (oxygen present). The self-purification of a stream is the result of the following separate natural occurrences that, although distinct, tend to merge one into the other.

Zone of degradation. In this first zone, oxygen is depleted and not completely exhausted. At this point, visible evidence of pollution such as floating solids and bits of paper will be present.

Biological life is present in great quantities. Fungus growth is probably also present, characterized by stringy growths that cling to any fixed surface in the stream. This active growth of the microorganisms will deplete and eventually exhaust the dissolved oxygen. At this point, most fish life will have disappeared, and unless this is a fast-flowing stream, sludge will begin to deposit on the bottom and in quiescent areas along the banks.

Zone of decomposition. This zone exists as a result of complete oxygen depletion. Anaerobic decomposition and putrefaction begin. If the pollution load is heavy enough, this will occur almost instantly. Otherwise, this zone will proceed more slowly. If the volume of waste is relatively small, it is possible that the stream may not pass through this decomposition stage, entering instead the zone of recovery.

Zone of recovery. As the stream flows along absorbing oxygen from the atmosphere, dissolved oxygen in the water begins to appear in increasing amounts. The more turbulent the stream, the more oxygen absorbed, to the point where the water may even become saturated with oxygen. As this occurs, organic solids will decrease, and the water begins to appear cleaner. The microorganisms are reduced in number, and aerobic species will begin to exist. Meanwhile, the anaerobic solids and organisms will disappear. Fish can again begin to live in the water, and the sludge deposits that originally existed will begin to disappear.

Zone of clean water. As the zone of recovery continues, the stream will eventually pass into this latter stage. Oxygen is at the point of saturation, depending on factors such as the stream temperature.

The time required or the length of stream necessary for such purification depends on the amount of pollution, the amount of flow in the stream, the turbulence of flow, and the temperature of the water.

Degrees of Treatment
Unfortunately, each plant does not have its own private rapidly flowing stream into which it can dump its wastes. Fortunately, almost all plants in this country

have access to sewers. Nevertheless, it may be necessary to resort to some degree of treatment. These various treatments are classified as follows:

Preliminary treatment consists of one or a combination of the following:

● *Screening* - to remove large suspended matter that otherwise might plug pumps or damage subsequent equipment.

● *Grit removal* - to trap heavier solids not removed by a screen. This should be practiced in all treatment processes with either physical chemical treatment (PCT) or activated sludge.

● *Primary treatment* - This usually consists of screening or grit removal followed by sedimentation tanks of various configurations.

Normally, the preliminary treatment will reduce the pollution by anywhere from 10% to 35%, depending upon the character of the original waste. In the case of another waste, wherein most of the pollution is soluble, the lower efficiency rate would be expected.

Secondary treatment. Where one of the activated sludge processes is used for secondary treatment, the reduction in pollution load will be approximately 80% to 85%. In the case of PCT, secondary treatment can consist of the coagulation and sedimentation phase. The expected pollution load decrease will be approximately 50% to 70%.

Tertiary treatment. The semantics used here convey several possibilities. This phase can mean simple filtration of the secondary effluent; or it can mean the use of coagulation, settling, activated carbon, and filtration. Depending on which process is used, the removal of polluting substances should be in the range of 90% to 95%.

It is not likely that complete treatment will be necessary before discharge, but at least some degree of pretreatment is a must.

Beverage Plant Waste

A carbonated beverage plant can be used as an illustration of various categories into which total waste can be broken down. Some wastes contain little more than suspended solids. These include the following:

● Final rinse waters, which are perhaps the greatest single wastewater flow

● Backwash water (which in today's plant is generally a considerable amount) resulting from the filters and purifiers

● Cooling waters from equipment such as compressors

● Wastes resulting from water treatment

There is nothing new about the recovery and reuse of these waters, only that the need to do so was never as necessary as it is today.

Recovery of backwash water has been practiced for at least 20 years. The reason, however, was not recovery, but to give assurance that an uncontaminated source or at least a treated source of backwash water was available. Today, of course, since the quantities are so much greater, this water can be recovered as a matter of economy. As to the recovery of final rinse waters, this too is not new, as indicated in a paper presented at a meeting of the Society of Soft Drink Technologists (SSDT) in 1964. The information contained therein (see Chapter XV) sounds better today than it did at that time.

In addition to these mildly polluting and recoverable flows from a carbonated beverage plant, there are flows that contain a high degree of pollution. These are the wastes that will require the utmost attention.

This discussion will concern itself only with the qualitative character of the waste. Following are the wastes that cause pollution:

● Pre-rinse water

● Wastes from syrup storage

● Wastes from syrup preparation

● Caustic wastes from soaker

● Washup wastes

● Truck-washing wastes

● Sanitary sewage

Figure 18-1 is a flow diagram of the points from which the pollution will emanate.

Wastewater chemistry is an extension of basic water chemistry. Both involve the analysis and treatment of liquid that contains polluting impurities. In water treatment, the inorganic chemistry involved is quite straightforward. In waste treatment, some knowledge of organic chemistry is required.

The complexity of organic formulae or reactions will not be necessary for the purposes of this discussion. Some of the terminology and their definitions will suffice. The terms most often used are *organics,* biological (or biochemical) oxygen demand *(BOD)*, and chemical oxygen demand *(COD)*.

Organic compounds contain carbon combined in one form or another with one or more other elements such as hydrogen, oxygen, and nitrogen. Most organic compounds are derived from these sources:

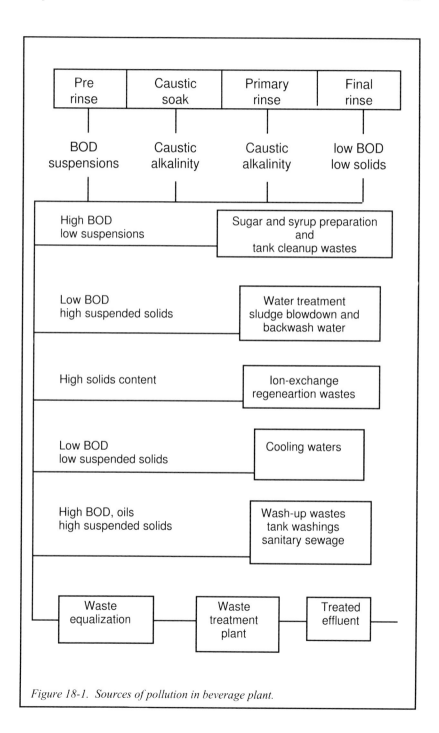

Figure 18-1. Sources of pollution in beverage plant.

● Nature — in compounds such as sugar, starch, fat, and oil.

● Fermentation — compounds such as alcohol and glycerin.

● Man-made — many organic compounds in use today are synthetic.

Organic compounds differ from inorganic compounds basically in that they serve as a source of bacteria. This is the reason for getting rid of them. It is well known what happens in a beverage plant when there is a source of bacteria, and the same holds true for a receiving body of water.

BOD and COD are the measure of pollution. Note that both terms contain the letter O, which is the chemical symbol for oxygen, the basis of biological waste treatment.

Biological (or biochemical) oxygen demand is that quantity of oxygen used up by bacteria in consuming the organic matter in a sample of wastewater at a temperature of 20 °C over a 5-day period. This is the basic test for pollution since it indicates results in terms of dissolved oxygen that would be consumed if the untreated waste were to be discharged into a natural body of water.

As long as a body of water contains free oxygen for fish and plant life to consume, it is not polluted (assuming no other toxic chemicals are present).

Complete biological oxidation takes place in 20 days. This much time usually cannot be taken for tests, however, so the 5-day BOD is used. The 5-day BOD is usually about 66% to 75% of the 20-day BOD value.

Chemical oxygen demand is a measure not only of the oxygen used by bacteria, but also of the oxygen used up by other substances that can be oxidized (e.g., iron and manganese). In other words, it gives the value of the total oxygen a waste can take from the natural body of water.

The main advantage of the COD value is that it can be determined in about 3 hours. Since each waste has a fairly constant ratio between BOD and COD, close BOD estimates can be made from a COD determination. Both COD and BOD are reported in ppm or gm/L.

The magnitude of beverage plant waste. It is possible to obtain some idea of the magnitude of pollution imparted by the waste from a carbonated beverage plant. For comparison, sanitary sewage has a pH close to 7, and contains about 1.6 lb of suspended solids per capita per day. It has a BOD of about 200 ppm (1.67 lb per capita per day).

Waste from a beverage plant will have a pH of between 8 and 11, with really no limit on suspended solids. It has a BOD of somewhere between 200 and 700 ppm, with an average of about 500 ppm (4.14 lb). Thus it can be seen that the plant discharge is approximately 2 to 3 times as strong or polluting as human waste.

The goal in waste treatment is to remove oxygen-consuming materials (whether they are inorganic or organic), so that the treated waste has an oxygen residual of some 5 to 8 ppm.

The wastewater from a beverage plant is alkaline, usually with a pH greater than that acceptable for discharge to a municipal waste treatment process or even into a waterway. Therefore, the first step in any disposal method is to neutralize this high-pH water to a pH value between 6 and 9 (EPA requirement). This can be performed through the use of mineral acids such as hydrochloric or sulfuric, or perhaps more simply through the use of carbon dioxide (CO_2). (See discussion later in this chapter).

Methods of Treatment

Generally speaking, there are three classifications of treatment methods. These are physical chemical treatment, biological treatment, or a combination of the two.

Physical chemical treatment consists of coagulation, settling, filtration, and/or adsorption. Operators by now are familiar with the first three of the unit processes. Adsorption using granulated or powdered carbon is also a reliable and effective way of reducing organic impurities.

Carbon will absorb detergents, insecticides, and those pollutants found in the wastes from a beverage plant. In this process the waste is collected in a holding and equalization tank with a minimum capacity of 2 to 3 hours of flow. Chemical coagulant, such as those used in water treatment, will be added to the equalized waste. After coagulation, the flow is then sent to a clarifier where the suspended solids are removed. The clarifier effluent is then filtered through a sand filter.

Following filtration, all of the suspended or insoluble pollutants will have been removed. Much of the pollutant remains in solution, however, and this must be removed.

Powdered carbon can be added to the filter effluent. After a short mix and retention, this spent carbon is removed by either clarification or filtration.

It is possible to use beds of granular carbon, either gravity or pressure type, in the same manner in which water is dechlorinated.

Since activated carbon removes organic compounds that are not responsive to or are difficult to remove by clarification or biological means, the use of carbon will undoubtedly increase.

Biological treatment. Following equalization and clarification of the plant waste, it is necessary to further treat the wastes, since much of the pollutant is in soluble form. This further treatment employs biological growth to effect aerobic decomposition or oxidation of the organic matter into more stable compounds, thus providing the higher degree of treatment.

Microorganisms play the key role in biological destruction of the organic pollutant. The two biological systems that can be used, trickling filters and activated sludge, are really basically the same.

Trickling filters are not filters at all. A trickling filter is basically a pile of 3- to 4-inch-diameter rocks, stacked from 6 to 8 feet high, over which the waste is allowed to trickle (Figure 18-2 illustrates a typical gravity filter). These rocks simply provide a surface area on which microbes cling and grow while they feed on the organic matter in the waste.

A trickling filter is used as an intermediate step of a treatment plant.

Raw --> Primary --> Trickling --> Final --> Effluent
Water   Clarifier Filter   Clarifier

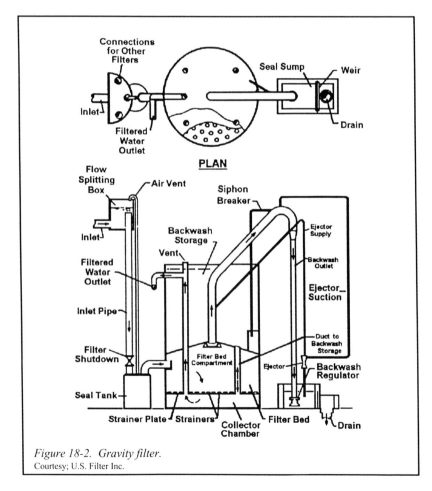

Figure 18-2. Gravity filter.
Courtesy; U.S. Filter Inc.

Activated sludge. In contrast to the fixed biological growth in the trickling filter, activated sludge plants consist of aerated basins in which the microbial growth is constantly agitated in the presence of air from which oxygen for the active aerobic process is derived.

Microorganisms play a very important part in the normal biological cycle of nature and, when properly controlled, can be used to treat sanitary and industrial wastes.

If a stream of polluted water were allowed to flow over a long-enough distance, it could probably cleanse itself. However, the pollution loadings that are added to most streams are now such that, unless man-made abatement plants are used, no aquatic life could exist in these streams.

One of these man-made remedies is the utilization of certain microorganisms to convert the oxidizable wastes and break them down into simple chemicals (carbon dioxide, water, sulfates, and nitrates). These organisms are also effective in conversion of substances to settleable solids.

The biological process can be achieved by two means. One is a low-rate process such as an aerated pond that uses oxidation only. The other is the activated sludge process in which the biochemical oxidation is carried out by

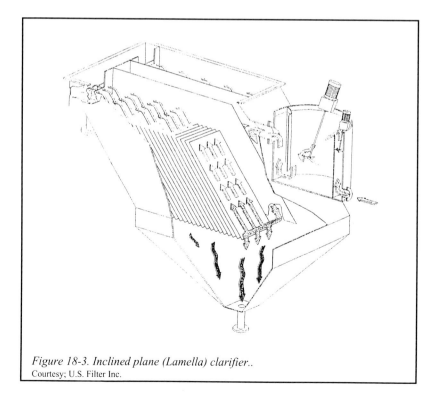

Figure 18-3. Inclined plane (Lamella) clarifier..
Courtesy; U.S. Filter Inc.

living organisms.

The term *activated sludge process* refers to a method of wastewater treatment. *Activated sludge* refers to the particles that are produced in wastewater by the growth of microorganisms in the presence of dissolved oxygen. The term *activated* comes from the fact that such a culture abounds in bacteria, fungi, and protozoa.

The activated sludge process is used to convert nonsettleable substances, which are in finely divided, colloidal, and dissolved forms, into settleable sludge and to remove this sludge. The sludge removal is accomplished in clarification tanks.

The activated sludge process depends on groups of microorganisms, primarily bacteria and protozoa living on wastewater solids, as a purifying medium. These organisms are maintained in an aerobic environment by introducing air into a mixture of activated sludge and wastewater in the aeration basins. Activated sludge is a brownish sludge floc consisting largely of organic matter obtained from the sewage, which is inhabited by bacteria and other forms of biological life. Activated sludge with its living organisms has the ability to absorb colloidal and dissolved matter from sewage so that the suspended solids are reduced. The biologic organisms use the absorbed material as food.

The basic sludge process is a versatile biological treatment that can be tailored to meet a wide variety of wastewater and effluent requirements. There are several general modifications of this system: step aeration, contact stabilization, and extended aeration.

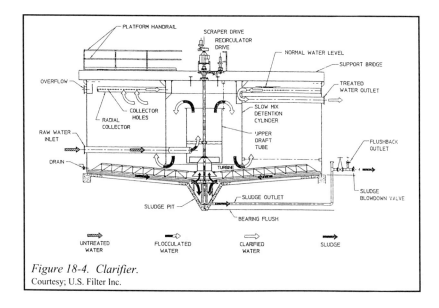

Figure 18-4. Clarifier.
Courtesy; U.S. Filter Inc.

Activated Sludge Modification
The basic system aerates the effluent from the primary clarifier to form the activated sludge. This sludge will consume the organic pollutants. Following aeration, the activated sludge mass flocculates and settles out in the secondary clarifier, leaving a clean effluent that is low in organics.

Pretreatment, also termed *preliminary treatment,* consists of the use of a screen and/or comminutor. A *comminutor* is a rotating device that intercepts and shreds the larger solids carried in the waste flow. It is necessary to maintain such solids in smaller form to avoid clogging of equipment downstream. Further, the treatment efficiency will be enhanced since the smaller the particle, the more contact there will be between the sludge particles and the air in the aeration basin.

Clarifier. The comminuted water flows into the clarifier (Figure 18-3 and 18-4), which is designed to afford quiescent settling of the suspended particles as the waste flows through the unit. Solids settle to the bottom of the clarifier tank. Because of the slow rotation of the scraper mechanism, these solids are concentrated and scraped into the sludge sump. These solids from a primary clarifier will be sent to a final disposal. When the clarifier is used to separate activated sludge particles from the flow, these recovered solids are called *return sludge.* The retention time of such clarifiers is between 2 and 3 hours at the average flow.

Figure 18-5. Aeration basin with a surface aerator.
Courtesy; Infilco Degremont Inc.

The clarifier will play a very critical role in any beverage plant biological waste treatment. Because this particular waste is high in carbohydrate content, it is prone to growth of a filamentous nature. Such growths require very low rise rates. Consequently, it is wise to design the clarifiers at a much slower speed than normal.

Aeration basins are usually rectangular basins with a retention time of between 2 and 8 hours, depending on the process used. Figure 18-5 shows a mechanical *surface aerator*, which is used to drive atmospheric air into the mixture of waste and return sludge. It is from the shearing action of this rotating aerator that oxygen from the air is transferred to the waste.

A *turbine aerator* is similar, except that the aerator is submerged and the air to be used is furnished by a blower. Another aerator is a *diffused air system,* in which compressed air is bubbled through various porous media.

The generation of activated sludge in sewage is a slow process. The amount formed from any volume of sewage during its period of treatment is small, and inadequate for the rapid and effective treatment of the waste that requires large concentrations of activated sludge.

Such concentration is built up by collecting the sludge treated and reusing this returned sludge in the treatment of subsequent sewage flows. The concentration of activated sludge will continue to increase until more sludge has been produced than is needed. The surplus is then removed from the treatment process and is known as *waste sludge* or *excess sludge.*

Step aeration process. As shown in Figure 18-6, the flow of settled wastes from the primary clarifier is split in the step aeration process so that it is distributed

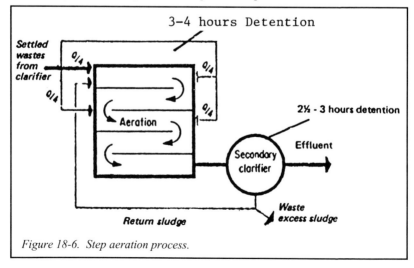

Figure 18-6. Step aeration process.

evenly between the sections of the aeration tank. Overall efficiency removal is increased since the oxygen demand from the beginning to end is spread throughout the system.

Contact stabilization process. In this modification of the basic system, the raw waste is mixed with aerated return sludge and the mixture is then aerated. However, the main supply of oxygen is added in the contact aeration chamber (see Figure 18-7). This heavy aeration renews and increases the adsorptive quality of the return sludge.

Extended aeration process. In this method, shown in Figure 18-8, the wastewater is handled without primary sedimentation. Following the use of grit removal and comminution, the wastewater flows directly into the aeration basin, which should have a detention time of *not less than 24 hours*. The aeration basin in this process serves the very useful purpose of acting also as the equalization basin to smooth out variations in load and flow, and has the capacity to dilute sludge-concentrated or toxic impurities in the wastewater. This activated sludge process consists of the following steps:

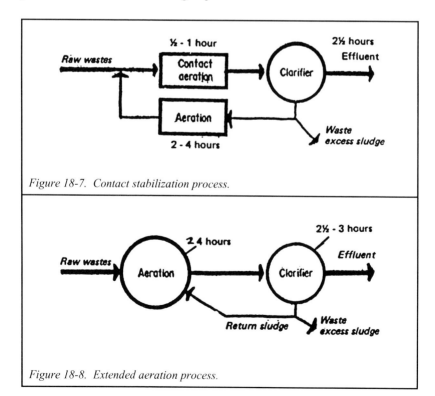

Figure 18-7. Contact stabilization process.

Figure 18-8. Extended aeration process.

● Mixing the activated sludge with the sewage being treated. The mixture of sewage and activated sludge in the aeration basin is known as *mixed liquor.*

● Aeration and agitation of this mixed liquor. (These are performed through the use of diffusers or mechanical aerators.)

● Separation of the activated sludge from the mixed liquor. (This is carried out in a clarifier.)

● Return of the proper amount of activated sludge for mixture with the incoming sewage.

● Disposal of the excess sludge to the sludge holding tanks and finally to the sludge drying beds.

The degree of treatment effected in extended aeration is the greatest available from any of the activated sludge modifications except for the combined treatment described below. Usually an extended aeration system will reduce the pollution by some 90% to 95%. This system is the simplest to control and operate, but keep in mind that some 24 hours of aeration plus about 3 to 4 hours of clarification time are needed.

A further advantage of this type of system is that the resulting solids are minimal, highly mineralized, and fairly easy to get rid of. Normally these solids can be used as landfill and, because of their high degree of mineralization, they will not cause any offensive odors.

Combined treatment. This is nothing more than the filtration of one of the above effluents and its treatment with carbon as described previously. This treatment would result in BOD reductions of greater than 90% to 95%. In fact, with proper chlorination techniques and control, this could result in a potable water.

Excess sludge treatment. In each of the processes previously discussed, approximately one-half pound of solids is formed for each pound of soluble BOD that is removed from the waste. Also in each of the processes, something less than this one-half pound of solids is returned to the aeration basin. As a result, more sludge is produced than is required to maintain an efficient oxidation of the incoming raw waste. That portion beyond the requirement is termed *excess sludge.*

The amount of excess sludge produced in the activated sludge processes is dependent upon the particular variation of the basic process. A high-rate activated sludge process produces the most excess sludge. A conventional activated sludge process produces less excess sludge. In general, the longer the retention time in the aeration basin, the smaller the amount of excess sludge produced. As a consequence, with its minimum 24-hour aeration basin, the extended aeration produces the least sludge. Theoretically, the extended aeration process should produce no excess sludge; hence, it is also known as a *total oxidation process*. Nevertheless, experience has shown that it does produce a certain amount of excess sludge.

Since any excess sludge contains all that remains of the original pollution that was generated (in this case, by the beverage plant), some means must be devised to remove it from the environment.

As this excess sludge is discharged from the final clarifier, it has a concentration of perhaps from 2% to 3% on a dry weight basis. This then would mean that this sludge still contains from 97% to 98% water. Some of the sludge solids still will contain organic contamination.

If there is sufficient land available near the plant, this liquid-solids mixture can be sent to a sand drying bed. Such a bed is somewhat similar to a filter bed. In this application, the liquid portion of the sludge is allowed to penetrate the sand bed and flow into the earth underneath. The water in the solids retained on the surface of the sand bed is then evaporated into the atmosphere until the remaining residue on the sand bed filter surface can be scraped off and disposed of as landfill. This would perhaps be the simplest way to remove the excess solids, but the necessary land is usually not available.

In almost every instance, it will probably be necessary to send this excess sludge to an aerobic digester. In such a system, the sludge is aerated very heavily in order to further and finally oxidize any remaining organic matter. The supernatant is drawn off and returned to the head end of the treatment process. The concentrated solids can then be disposed of by various methods such as disposal to the local sewage treatment plant or use as landfill.

A Summary of Wastewater Treatment Processes
The treatment of wastewaters falls into three categories, as summarized below and as shown in the flowchart in Figure 18-9.

- *Primary* — the removal of suspended solids.

- *Secondary* — the removal of biodegradable substances. This means the removal of all substances that can be altered and thus removed by biological means or oxidation using oxygen or ozone.

- *Tertiary* — this usually refers to additional treatment of the effluent from secondary treatment. As such, it can be considered as the "polishing" and ultimate treatment of wastewater, whether municipal or industrial

The Use of Ozone (O_3) in Wastewater Treatment.
In the discussion of activated sludge processes, it was noted that the oxygen required for the processes has been obtained from atmospheric air. Although this

is historically true, research and new techniques have led to the use of manufactured oxygen (O_2) to replace atmospheric air. During the past 10 years, a number of secondary treatment plants have been constructed that use pure oxygen manufactured through a cryogenic (low-temperature) process.

In a similar manner, the use of ozone (O_3) in the treatment of secondary wastes has been researched and even tried, with promise of better things to come. It has been surmised that since the basis of this method of waste treatment depends upon the use of an oxidizing agent, the more powerful the oxidant, the more efficient the process. Hence, since ozone (O_3) is more powerful than oxygen (O_2), it will produce more efficiently (see Chapter 12).

PRIMARY TREATMENT

Screening	Grit removal	Primary clarification
	chlorination	
	35% BOD removal	

SECONDARY TREATMENT

Primary clarification	Chemical treatment	Secondary clarification
	Chlorination	
	65% BOD removal	

Primary clarification	Biological treatment	Secondary clarification
	Chlorination	
	80-90% BOD removal	

TERTIARY TREATMENT

Secondary treatment	Filtration	Clarification
	Ozonation	
	95% BOD removal	

EXTENDED AERATION

Comminution	Biological treatment	24 hour aeration
3-4 hour settling	Chlorination and/or	Ozonation
	95% BOD removal	

Figure 18-9. Schemes for waste treatment.

Ozone has been extensively used in other parts of the world in the treatment of water, and used also for removal of a number of substances from industrial wastes. It should come as no surprise that its application would eventually find its way to the treatment of beverage plant wastes.

The ozonator shown in Figure 18-10 has a modular design. The reason for this is economics. The use of a module reduces engineering design costs. The unit is sized depending upon two factors, the ozone demand and the rate of flow. The stronger the waste, the greater the ozone demand will be.

The rate of flow to the treatment plant will determine the size of the settling tanks. It will very likely be necessary to install a holding or equalization basin ahead of this ozonator to smooth out both the flowrate and the character of the waste. Since the flows from a beverage plant are sporadic throughout the day, the treatment plant likely will be overloaded at times and not flowing at other times. This matter of flow and loading should be carefully studied and engineered prior to the installation of *any* waste treatment plant.

Further, it may be required to condition the plant waste before treatment, such as with carbon dioxide for pH adjustment (see last section of this chapter).

Regardless of the source of ozone, the benefits of such a system should not be disregarded. There is no doubt as to its efficacy for reduction of organics, pollution, and odor; for readily disposal sludge; and for its absence of chlorinated compounds.

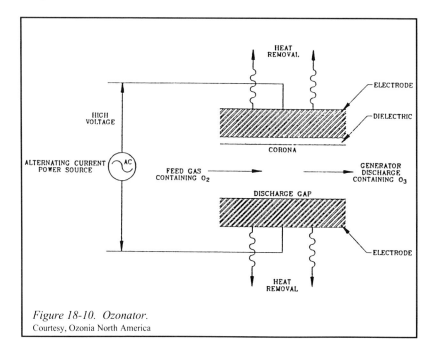

Figure 18-10. Ozonator.
Courtesy, Ozonia North America

At the same time, with an ozone system there are safety precautions to be considered. A unit such as the one shown in Figure 18-10 is hermetically sealed because excess or surplus ozone that is not used in the process is returned to the head end of the process. There is a general impression that ozone is an explosive hazard. However, experts agree that unless heavy concentrations (25 to 30 ppm, which are not likely) are in the presence of a volatile hydrocarbon (such as gasoline), and a spark is possible, it is highly unlikely that ozone will explode or burn. However, it behooves the user to make certain that this combination does not exist.

Additionally, ozone is toxic and dangerous to human life. Safeguards such as controlled emission into the atmosphere must be used. Devices such as automatic shutoffs should be designed into ozone-producing and ozone-using equipment.

The key factors in the use of such processes will be the proper engineering for each application, and the required experience in operation.

Neutralization of Alkaline Wastes
The wastewater from a bottling plant is alkaline, usually with a pH greater than would eventually be acceptable to a municipal treatment plant, waste treatment plant, waste treatment process, or even to a public waterway. Therefore, the first step in any of the disposal methods will be to neutralize this flow (pH 9 to 12) to a pH of approximately 7. (The EPA requirement is a pH of 6 to 9.)

If this plant flow is to be discharged to a waterway, the nearer it is to pH 7

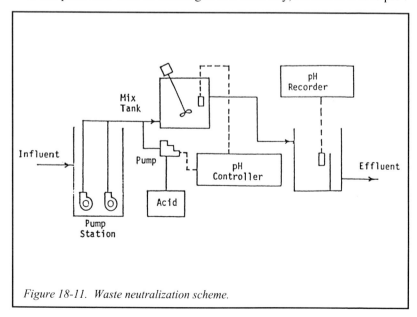

Figure 18-11. Waste neutralization scheme.

(neutral), the better. If the plant flow is sent to either a municipal sewer system, or if the beverage plant is to have its own waste treatment process, this neutralization should likewise be practiced. The more neutral this waste as it enters any waste treatment plant, the more operable and efficient will be the plant. This is especially true when it is considered that almost every municipal sewer treatment plant is of the activated sludge type and thus will require pH control. One reason for the EPA requirement is that if industry is made to pretreat at least to this extent, the pollution loadings sent to municipal treatment plants will be alleviated to this degree.

A typical waste neutralization scheme is shown in Figure 18-11. This neutralization can be carried out through the use of mineral acids such as hydrochloric (HCl) or sulfuric (H_2SO_4). It can also be done, perhaps more simply, with carbon dioxide (CO_2).

If HCl is used to neutralize this flow, which contains combinations of sodium hydroxide (NaOH), lime ($Ca(OH)_2$), and calcium carbonate ($CaCO_3$), the following reactions will result:

$$NaOH + HCl ----> NaCl + H_2O$$
$$Ca(OH)_2 + 2HCl ----> CaCl_2 + 2H_2O$$
$$CaCO_3 + 2HCl ----> CaCl_2 + H_2O + CO_2$$

Note that in each case the end products contain chlorides, and the more concentrated these alkaline wastes, the greater the resultant chlorides. On the other hand, if CO_2 is used for neutralization, the following reactions occur: First,

$$CO_2 + H_2O ----> H_2CO_3$$

Then,

$$NaOH + H_2CO_3 ---> NaHCO_3 + H_2O$$
$$Ca(OH)_2 + H_2CO_3 ----> CaCO_3 + H_2O$$
$$Ca(OH)_2 + 2H_2CO_3 ----> Ca(HCO_3)_2 + 2H_2O$$
$$CaCO_3 + H_2CO_3 ----> Ca(HCO_3)_2$$

The end products (carbonates) are certainly less objectionable than the mineral salts.

Applying the chemistry of equivalents as described in Chapters 4, 5, and 6, it can be determined that 44.01 pounds of CO_2 (100%) will neutralize the same amount of alkalinity as 36.465 lb of HCl (100%).

Molecular Weight CO_2	*Molecular Weight HCl*
C = 12.01	H = 1.0
2O = 32.00	Cl = 35.465
44.01	36.465

It has been determined that 4.0 lb of CO_2 are required to neutralize 1,000 gallons of alkaline wastewater (pH 12) to
pH 7. Therefore, it would require:

$$4.0 \text{ lb} \times (36.465) \div 44.1 = 3.3142 \text{ lb} (100\% \text{ HCl})$$

to neutralize the same quantity of wastewater. Since the acid used for this purpose would be diluted to perhaps 30%, the following would be required:

$$3.3142 \times 100\% \div 30\% = 11.047 \text{ lb} (30\% \text{ HCl})$$

or if a 50% solution is used:

$$3.3142 \times 100\% \div 50\% = 6.62 \text{ lb} (50\% \text{ HCl})$$

Regardless of the comparative costs, however, there appear to be these advantages in the use of CO_2:

● Carbonic acid cannot exist at a pH lower than 5.8, which makes this neutralization self-limiting; and because of this property:

● An overdose of CO_2 will not leave any residual acid in the system.

● Storage problems are greatly minimized and the handling and use of very corrosive acids is eliminated.

● The CO_2 is essentially inert, while the acids are corrosive and potential dangers lie in their use.❏

TABLES AND CONVERSIONS

Molecular Weight and CaCO₃ Conversion Factors for Ions in Water Treatment

Ion/Chemical	Chemical Formula	Molecular Weight	Substance to CaCO₃ Equiv.	CaCO₃ Equiv. to Substance Factor
Acetic Acid	$HC_2H_3O_2$	60.1	0.83	1.20
Aluminum	Al^{+++}	27.0	5.56	0.18
Ammonia	NH_3	17.0	2.94	0.34
Ammonium	NH_4^+	18.0	2.78	0.86
Barium	Ba^{++}	137.4	0.73	1.37
Bicarbonate	HCO_3^-	61.0	0.82	1.22
Calcium	Ca^{++}	40.1	2.50	0.4
Carbon Dioxide	CO_2	44.0	1.14	0.88
Carbonate	$CO_3^=$	60.0	0.83	1.20
Chloride	Cl^-	35.5	1.41	0.71
Chlorine	Cl_2	70.0	1.41	0.71
Copper	Cu^{++}	63.6	1.57	0.64
Iron (Ferric)	Fe^{+++}	55.8	2.69	0.37
Iron (Ferrous)	Fe^{++}	55.8	1.79	0.56
Fluoride	F^-	19.0	2.63	0.38
Hydrogen	H^+	1.0	50.0	0.02
Hydrogen Peroxide	H_2O_2	34.0	2.94	0.34
Hydrogen Sulfide	H_2S	34.1	2.93	0.34
Hydroxide	OH^-	17.0	2.94	0.34
Iodide	I^-	126.9	0.39	2.54
Iodine	I_2	253.8	0.39	2.54
Lead	Pb^{++}	207.	0.48	2.08
Magnesium	Mg^{++}	24.3	4.10	0.24
Manganese (Manganic)	Mn^{+++}	54.9	2.73	0.37
Manganese (Manganous)	Mn^{++}	54.9	1.82	0.55
Nitrate	NO_3	62.0	0.81	1.24
Peracetic Acid	$HC_2H_3O_3$	76.1	0.66	1.52
Phosphate	PO_4^{-3}	95.0	1.58	0.63
Potassium	K^+	39.1	1.28	0.78
Silica	SiO_2	60.1	0.83	1.20
Sodium	Na^+	23.0	2.18	0.46
Sulfate	SO_4	96.1	1.04	0.96
Sulfide	$S^=$	32.1	3.13	0.32
Water	H_2O	18.0	5.56	0.18

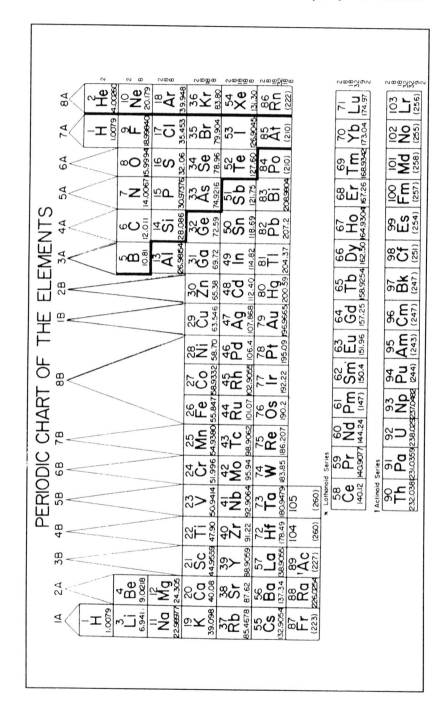

PERIODIC CHART OF THE ELEMENTS

Conversion of U.S. to Other Units

Units	Parts CaCO$_3$ per million (ppm)	Grains CaCO$_3$ per U.S. Gallon (grpg)	English degrees or Clark	French degrees	German degrees	Milliequivalents per liter meq/million
1 part per million	1.	0.0583	0.07	0.1	0.0560	0.20
1 Grain per U.S. Gallon	17.1	1.	1.2	1.71	0.958	0.343
English or Clark degree	14.3	0.833	1.	1.43	0.800	0.286
1 French degree	10.	0.583	0.7	1.	0.560	0.20
1 German degree	17.9	1.04	1.24	1.79	1.	0.357
1 Milliequivalent/L	50.0	2.92	3.50		2.80	1.

Conversion Units and Equivalents

Water analysis units	Parts per million (ppm)	Milligrams per liter (mg/L)	Grams per liter (g/L)	Grains U.S. gallon (grs/U.S. gal)	Grains British Imp. gallon	Kilograins per cubic foot (Kgr/ft^3)
1 Part per million	1.	1.	0.001	0.583	0.07	0.0004
1 Milligram per liter	1.	1.	0.001	0.583	0.07	0.0004
1 Gram per liter	1000.	1000.	1.	68.3	70.0	0.436
1 Grain per U.S. gallon	17.1	17.1	0.017	1.	1.2	0.0075
1 Grain per British Imp. gal	14.3	14.3	0.014	0.833	1.	0.0062
1 Kilograin per cubic foot	2294.	2294.	229.4	134.	1.	1.

1.0 milligram/Liter (mg/L) = 1.0 part per million 10^{-3} g/L
1.0 nanogram/Liter (ng/L) = 1.0 part per trillion 10^{-9} g/L
1.0 microgram/Liter (#g/L) = 1.0 part per billion 10^{-6} g/L

Useful Calculations.

1 gallon of water weighs approximately 8.33 pounds.
1 gallon = 231 cubic inches
1 gallon = 3.785 liters
1 gallon = 3785 milliliters
1 cubic foot of water weighs approximately 62.4 pounds
1 cubic foot = 7.48 gallons
1 pound = 7000 grains
1 pound = 453.6 grams
1 gram = 15.43 grains
1 gram per liter = 58.41 grains per gallon
1 gram per liter = 1000 ppm
0.0171 grams per liter = 1 grain per gallon
0.0038 grams per gallon = 1 ppm
1 grain per gallon = 17.1 ppm
1 grain per gallon = 142.9# per million gallons
1 milligram per liter = 1 ppm
1 ppm = 0.058 grains per gallon
8.33 pounds/1,000,000 gallons = 1 ppm
10,000 ppm - 1%

To Convert from	to	Multiply by
Capacity		
Kgrs/ft.³ (as CaCO₃)	g CaO/l	1.28
Kgrs/ft.³ (as CaCO₃)	g CaCO₃/l	2.29
Kgrs/ft.³ (as CaCO₃)	eq/l	0.0458
g CaCO₃ l	Kgrs/ft³ (as CaCO₃)	0.436
g CaO/l	Kgrs/ft³ (as CaCO₃)	0.780
Flow Rate		
U.S. gpm/ft.³	BV/hr	8.02
U.S. gpm/ft.²	m/hr	2.45
U.S. gpm	m³/hr	.227
BV/min	U.S. gpm/ft³	7.46
Pressure Drop		
psi/ft.	mH₂0/m resin	2.30
	g/cm²/m	230
Regenerant Concentration		
lbs/ft.³	g/l	16.0
Density		
lbs/ft.³	g/l	16.0
Rinse Requirements		
U.S. gal/ft.³	BV	0.134

COMMON CHEMICALS USED TO TREAT WATER

CHEMICAL	COMMON NAME	TYPICAL SPECS	EQUIV. WEIGHT	BULK DENSITY lb/cu ft or lb/gal	APPROX. ph 1% SOLUTION
Aluminum Sulfate $Al_2(SO_4)_3 \cdot 14H_2O$	Alum	Lump — 17% Al_2O_3 Liquid — 8.5% Al_2O_3	100[1]	60 11	3.4
Bentonitic Clay	Bentonite	—	—	60	—
Calcium Carbonate $CaCO_3$	Limestone	96% $CaCO_3$	50	80	9
Calcium Hydroxide $Ca(OH)_2$	Hydrated Lime, Slaked Lime	96% $Ca(OH)_2$	40[1]	40	12
Calcium Hypochlorite $Ca(OCl) \cdot 4H_2O$	HTH	70% Cl_2	103	55	6—8
Calcium Oxide CaO ,,	Burned Lime, Quicklime	96% CaO	30[1]	60	12
Calcium Sulfate $CaSO_4 \cdot 2H_2O$	Gypsum	98% Gypsum	86[1]	55	5—6
Chlorine (Cl_2)	Chlorine	Gas — 99.8% Cl_2	35.5	gas	—
Copper Sulfate $CuSO_4 \cdot 5H_2O$	Blue Vitriol	98% Pure	121[1]	75	5—6
Dolomitic Lime $Ca(OH)_2 \cdot MgO$	Dolomitic Lime	36—40% MgO	67[2]	40	12.4
Ferric Chloride $FeCl_3 \cdot 6H_2O$	Iron Chloride	Lump — 20% Fe Liquid — 20% Fe	91[1]	70 13	3—4
Ferric Sulfate $Fe_2(SO_4)_3 \cdot 3H_2O$	Iron Sulfate	18.5% Fe	51.5[1]	70	3—4
Ferrous Sulfate $FeSO_4 \cdot 7H_2O$	Copperas	20% Fe	139[1]	70	3—4
Hydrochloric Acid HCl	Muriatic Acid	30% HCl 20° Baume	120[1]	9.6	1—2
Sodium Aluminate $NaAlO_2$	Aluminate	Flake — 46% Al_2O_3 Liquid — 26% Al_2O_3	100[1]	50 13	11—12
Sodium Chloride NaCl	Rock Salt, Salt	98% Pure	58.5	60	6—8
Sodium Carbonate Na_2CO_3	Soda Ash	98% Pure 58% Na_2O	53	60	11
Sodium Hydroxide NaOH	Caustic, Lye	Flake — 99% NaOH Liquid — 50—70%	40	65 12	12.8
Sodium Phosphate Na_2HPO_4	Disodium Phosphate	49% P_2O_5	47.3	55	9
Sodium Metaphosphate $NaPO_3$	Hexameta-phosphate	66% P_2O_5	34	47	5—6
Sulfuric Acid H_2SO_4	Oil of Vitriol	94—96% 66° Baume	50[1]	15	1—2

(1) Effective equivalent weight of commercial product.
(2) Effective equivalent weight based on Ca(OH)$_2$ content.

Courtesy National Lime Association, Washington, D.C.

GLOSSARY

A

absorb - to ingest, to swallow.

absorption - assimilation of molecules or other substances into the physical structure of a liquid or solid without chemical reaction.

acid - the hydrogen form of an anion.

activated sludge - sludge floc produced in raw or settled waste by the growth of bacteria and other organisms in the presence of oxygen.

adsorb - to collect on a surface.

adsorption - physical adhesion of molecules or colloids to the surfaces of solids without chemical reaction.

aerobic organism - an organism that requires oxygen for its survival.

agglomerate - to gather fine particles into a larger mass.

algae - primitive chlorophyll-containing plants, one- or many-celled, usually aquatic, and capable of elaborating their foodstuffs by photosynthesis. Many are microscopic, but under conditions favorable for their growth, they grow in colonies and produce mats and similar nuisance masses.

algicide - any substance that kills algae.

alkalinity - (abbr. alk) - bicarbonate, carbonate, or hydrate amounts in water. Can be expressed as "M" alk to a methyl orange titration end point (about pH 4.2) or "P" alk to a phenolphthalein end point (about pH 8.2).

anaerobic organism - one that can survive in the absence of oxygen.

anion - a negatively charged ion resulting from dissociation of salts, acids, or alkalis in aqueous solution, and attracted to the anode.

atom - the smallest part of an element that can take part in a chemical reaction without being permanently changed.

B

bacteria - primitive plants, generally free of pigment, which reproduce by dividing in one, two, or three planes. They occur as single cells, groups, chains, or filaments, and do not require light for their life processes.
- microscopic, single-celled plants that reproduce by fission or by spores, identified by their shapes: coccus, spherical; bacillus, rod-shaped; and spirillum, curved.

bacterial count - a measure of the concentration of bacteria.

biochemical - resulting from biological growth or activity, and measured by or expressed in terms of the ensuing chemical change.

biochemical action - chemical changes resulting from the metabolism of living organisms.

biochemical oxygen demand (BOD) (also known as *biological* oxygen demand) - the quantity of oxygen utilized in the biochemical oxidation of organic matter, measured under controlled test conditions (in a specified time and at a specified temperature). It is not related to the oxygen requirements in chemical combustion, being determined entirely by the availability of the material as a biological food and by the amount of oxygen utilized by the organism during oxidation.

biochemical oxygen demand, standard - biochemical oxygen demand as determined under standard laboratory procedure for 5 days at 20 °C, usually expressed in parts per million.

biota - all living organisms of a region or system.

C

carbonate hardness - that hardness in a water caused by bicarbonates and carbonates of calcium and magnesium. If alkalinity exceed total hardness, all hardness is carbonate hardness; if hardness exceeds alkalinity, the carbonate hardness equals the alkalinity.

carcinogen - a substance that produces cancer.

cation - a positively charged ion, resulting from dissociation of molecules in solution, and attracted to the cathode.

caustic soda - a common water treatment chemical, sodium hydroxide; lye.

chelating agents - organic compounds having the ability to withdraw ions from their water solutions into soluble complexes.

chemical oxygen demand (COD) - a measure of organic matter and other reducing substances in water.

chloramines - compounds of organic amines or inorganic ammonia with chlorine.

chlorination - the application of chlorine.

>*available chlorine* - a term used in rating chlorine and hypochlorites as to their total oxidizing power.

>*breakpoint chlorination* - the application of chlorine to water or industrial wastes containing free ammonia, to provide free residual chlorination.

>*chlorine demand* - the difference between the amount of chlorine added to water or industrial wastes and the amount of residual chlorine remaining at the end of a specified contact period. The demand for any given water varies with the amount of chlorine applied, time of contact, and temperature.

>*combined available residual chlorine* - of the total residual chlorine remaining in water or industrial wastes at the end of a specified contact period, that portion that will react chemically and biologically as chloramines, or organic chloramines.

>*free available residual chlorine* - of the total residual chlorine remaining in water or industrial wastes at the end of a specified contact period, that portion that will react chemically and biologically as hypochlorous acid, hypochlorite ion, or molecular chlorine.

>*prechlorination* - the application of chlorine to water or industrial wastes prior to any treatment. The term refers only to a point of application.

>*postchlorination* - the application of chlorine to water or industrial wastes subsequent to any treatment. The term refers only to a point of application.

>*residual chlorine* - the total amount of chlorine (combined and free available chlorine) remaining in water or industrial wastes at the end of a specified contact period following chlorination.

coagulation - (1) the agglomeration of colloidal or finely divided suspended matter by the addition to the liquid of an appropriate chemical coagulant, by biological processes, or by other means. (2) the process of adding a coagulant and the necessary reacting chemicals, which brings together small particles into a single larger mass that can be filtered or settled out.

coliform bacteria - those found in the intestinal tracts of warm-blooded animals, and used as indicators of pollution if found in water.

colloids - finely divided solids that will not settle but may be removed by coagulation and by some but not all types of filtration. The matter is of very fine particle size, usually in the range of 10^{-5} to 10^{-7} cm in diameter.

communition - the process of screening waste flows and cutting the screenings into particles sufficiently fine to pass through the screen openings.

concentration - the process of increasing the dissolved solids per unit volume of solution, usually by evaporation of the liquid; also, the amount of material dissolved in a unit volume of solution.

condensate - water obtained by evaporation and subsequent condensation.

conductivity - ability of a substance to conduct heat or electricity; the ability of electric current to flow through water as a measure of its ion content in mhos or micromhos.

contaminant - foreign component present in another substance (e.g., anything in water that is not H_2O is a contaminant).

D

deaerator - a unit or system used to remove gases from water.

demineralization - process used to remove minerals from water; normally used to describe ion-exchange processes.

detention - the theoretical time required to displace the contents of a tank or unit at a given rate of discharge (volume divided by rate of discharge).

dialysis - a process used for separation of ions from water; the use of a certain membrane that is impermeable to water allows ions to pass through while retaining the water.

diatomaceous earth - a filter medium that is made up of the siliceous skeleton of organisms related to algae.

diatoms - organisms related to algae, having a brown pigmentation and a siliceous skeleton.

disinfection - the killing of the larger portion (but not necessarily all) of the harmful and objectionable microorganisms in, or on, a medium by such means as chemicals, heat, or ultraviolet light.

disinfection by-products (DBP) - compounds resulting from the reaction of a disinfectant and certain organic materials found in water (usually surface supplies).

E

efficiency - the ratio of the actual performance of a device to the theoretically perfect performance, usually expressed as a percentage.

> *average efficiency* - the efficiency of a machine or mechanical device over the range of load through which the machine operates.

> *filter efficiency* - the operating results from a filter as measured by various criteria such as percentage reduction in suspended matter, total solids, biochemical oxygen demand, bacteria, or color.

effluent - (1) a liquid that flows out of a containing space; (2) waste, water, or other liquid, partially or completely treated, or in its natural state, as the case may be, blowing out of a reservoir, basin, treatment plant, tank, or system.

electrodialysis - a form of dialysis that uses direct electric current as the driving force.

electrolyte - a substance that dissociates into two or more ions that it dissolves in water.

element - a substance that cannot be decomposed without losing the chemical or physical properties that make it unique.

equivalent weight or *equivalent, chemical* - the weight in grams of a substance that combines with or displaces 1 gram of hydrogen; usually obtained by dividing the formula weight by the valence.

ethylenediaminetetraacetic acid (EDTA) - The sodium salt is the usual form of this chelating material.

eutrophication - enrichment of water, causing excessive growth of aquatic plants and an eventual choking and deoxygenation of the water body.

exchange sites - reactive groups on an exchange resin.

F

facultative organisms - microbes capable of adapting to either aerobic or anaerobic environments.

filtrate - the effluent of a filter.

filtration - the process of separating solids from a liquid by means of a porous substance through which only the liquid passes.

floc - small gelatinous masses, formed in a liquid by the addition of coagulants or through biochemical processes or by agglomeration.

flocculation - the process of agglomerating coagulated particles into settleable flocs.

flocculator - an apparatus for the formation of floc in water or waste.

flotation - a process for separating solids from water by developing a froth in a vessel in such fashion that the solids attach to air particles and float to the surface for collection.

flume - a raceway or channel constructed to carry water or to permit measuring its flow.

freeboard - the space above the resin bed to accommodate resin expansion in backwash; also,the vertical distance between the normal maximum level of the surface of the liquid (as in a conduit, reservoir, tank, or canal); and the top of the sides of that containment. It is provided so that waves and other movements of the liquid will not overtop the confining structure.

fungi - as applied to water, these are simple, one-celled organisms without chlorophyll, often filamentous; molds and yeasts are included in this category.

G

germ - microscopic plant or animal that causes diseases.

grains per gallon (gr/gal) - 1 gr/gal = 17.1 mg/L.

H

hardness (total) - the concentration of calcium and magnesium salts in water. Hardness is a term originally referring to the soap-consuming power of water; as such, it is sometimes also taken to include iron, manganese, and other multivalent cations.

> *permanent hardness* - the difference between total and temporary hardnesses; the excess of hardness over alkalinity; also referred to as carbonate hardness.

> *temporary hardness* - the hardness equal to the sum of the bicarbonates, carbonates, and hydroxides of calcium and magnesium; hardness equal to or less than the alkalinity; noncarbonate hardness.

I

influent - waste, water, or other liquid, raw or partly treated, flowing into a reservoir, basin, or treatment plant.

inorganic matter - those substances not having the structure of living things (plants or animals).

ion - an atom or radical in solution carrying an integral electrical charge, either positive (cation) or negative (anion)

ion exchange - a process by which certain undesired ions of given charge are absorbed from solution within an ion-permeable absorbent, being replaced in the solution by desirable ions of similar charge from the absorbent.

J

jar tests - a laboratory method to determine the ideal dosages of chemical required to treat water or waste to the desired degree.

L

Langelier index - a means of expressing the degree of saturation of a water as related to calcium carbonate solubility. This value determines the point at which water is neither scale-producing or corrosive.

lignin - the major noncellulose constituent of wood.

lime - a common water-treatment chemical. Limestone, $CaCO_3$, is burned to produce quicklime, CaO; which is mixed with water to produce slaked (hydrated) lime, $Ca(OH)_2$.

M

membrane - a thin pliable material used as a barrier in various conformations as a filter medium. Different membranes permit the passage only of particles up to a certain size or of special nature, and thus are used for the removal of different sizes or natures of particles.

mho - unit of electrical conductance; reciprocal of an ohm.

micron (µm) - one one-millionth of a meter; 10^{-6} meter.

microorganism - minute organism either plant or animal, invisible or barely visible to the naked eye; must be viewed under a microscope, as compared to larger, visible types called *macroorganisms*.

mineral - a natural inorganic substance having a definite chemical composition and structure; neither plant nor animal.

molecule - a combination of atoms that results in a compound; the smallest particle of an element or compound retaining its characteristics.

N

nanometer - one one-billionth of a meter; 10^{-9} meter.

nitrilotriacetic acid (NTA) - a chelant with the sodium salt being the usual form.

noncarbonate hardness - hardness in water caused by chlorides, sulfates, and nitrates of calcium and magnesium.

noncondensables - gaseous material not liquefied when associated water vapor is condensed in the same environment.

nonpathogen - a microorganism or virus that does not cause disease.

nonvolatile organic compounds (NVOC) - substances that are not readily volatile.

O

occlusion - an absorption process by which one solid material adheres strongly to another, sometimes occurring by co-precipitation.

ohm - a unit of resistance to the passage of electrical current.

organic matter - those substances that have the characteristics of, or derive from, living organisms (plant or animal).

orifice - an opening through which a fluid can pass; a restriction placed in a pipe to provide a means to measure flow.

osmosis - the passage of water through a permeable membrane separating two solutions of different concentrations; the water passes into the more concentrated solution.

overflow rate - a criterion for the design of settling units; expressed as gallons per day per square foot (gpd/ft^2) of settling area in the clarifier.

oxidation - the addition of oxygen; the removal of hydrogen to result in the increase in valence of an atom; a chemical reaction in which an element or ion is increased in positive valence, losing electrons to an oxidizing agent.

P

parts per million (ppm) - milligrams per liter, expressing the concentration of a specified component. A ratio of pounds per million pounds, grams per million grams, etc.

pathogens - disease-producing microbes.

periodic chart - an arrangement of the elements in order of increasing atomic numbers, in a way to illustrate the repetition (or periodicity) of key characteristics.

pH - a means of expressing hydrogen ion concentration in terms of the powers of 10. This number system relates to the acidity, neutrality, or basicity of water.

photosynthesis - the process of converting carbon dioxide and water to carbohydrates, activated by sunlight in the presence of chlorophyll, liberating oxygen.

plankton - small organisms with limited powers of locomotion, carried by water currents from place to place.

pollutant - a contaminant at a concentration high enough to endanger the aquatic environment or the public health.

polymer - a chain of organic molecules produced by the joining of primary units called *monomers*.

polyphosphate - molecularly dehydrated ortho-phosphate.

potable - satisfactory for human consumption.

potable water - meets drinking-water quality standards.

precipitate - an insoluble reaction product of an aqueous chemical reaction, usually a crystalline compound that grows in size to become settleable.

protozoa - large microscopic single-cell organisms higher on the food chain than bacteria, which they consume.

purification - the removal, by natural or artificial methods, of objectionable matter from water.

R

reduction - opposite of oxidation; a chemical reaction in which elements or compounds gain electrons and thus reduce in positive valence.

reverse osmosis - The opposite of osmosis; a process that reverses the flow of water through a membrane by applying pressure to overcome the osmotic pressure, so that the water passes from the more concentrated to the more dilute solution.

S

salinity - the presence of soluble minerals in water.

saturation index - the relation of calcium carbonate to the pH, alkalinity, and hardness of a water to determine its scale-forming tendency.

scale - the precipitate that forms on surfaces in contact with water as a result of a physical or chemical change.

sedimentation - gravitational settling of solid particles in a liquid system.

sequester - to form a stable, water-soluble complex.

soda ash - a common water-treating chemical, sodium carbonate.

softening - the removal of hardness — calcium and magnesium — from water, replacing it with sodium by cation exchange.

spore - a reproductive cell, or seed, of algae, fungi, or protozoa.

stability index - an empirical modification of the saturation index used to predict scaling or corrosive tendencies in water systems.

synergism - the combined action of several chemicals that produces an effect greater than the additive effects of each.

T

thermocline - the layer in a lake dividing the upper, current-mixed zone from the cool lower, stagnant zone.

trihalomethane - a compound that results from the reaction of chlorine with certain organic compounds (found usually in surface supplies).

turbidity - a suspension in water of fine particles that obscure light rays but that require many days for sedimentation because of small particle size; also, a measure of suspended materials by optical obstruction of light rays passed through a water sample, as compared to a standard.

turnover - in impounded water supplies, such as a lake, the mixings of upper and lower layers of water; caused by temperature changes and density equalization that occur in spring and fall.

U

ultraviolet rays - light rays of extremely short length that can be adapted to the disinfection of water and waste.

V

volatile organic compounds (VOC) - compounds that are readily vaporized.

valence - the number of positive or negative charges of an ion.

W

weir - a dam with an edge or notch used for equalizing and/or measuring flow.

ACRONYMS and ABBREVIATIONS

BOD - biological oxygen demand

CIP - clean in place
cm³ - cubic centimeters
COD - chemical oxygen demand
CR - continuous regeneration

DBP - disinfection by-products
DC - direct current
DE - diatomaceous earth

ED - electrodialysis
EDR - electrodialysis reversal
EPA - Environmental Protection Agency

ft² - square feet
ft³ - cubic feet

g - grams
GAC - granular activated carbon
gal - gallons
gal/min/ft² - gallons per minute per square foot
gpd - gallons per day
gph - gallons per hour
gpm - gallons per minute
gr/ft³ - grains per cubic foot
gr/gal - grains per gallon
g/L - grams per liter

h - hour
hp - horsepower
HPC - heterotrophic plate count

in³ - cubic inches

J.T.U. - Jackson turbidity units

kgr/ft³ - kilograins per cubic foot

L - liters
lb - pounds
lb/gal - pounds per gallon
lb/h - pounds per hour
LSI - Langelier saturation index

MBAS - methylene blue active substance
MCL - maximum contaminant level
meq/L - milliequivalents per liter

MF - microfiltration
mg - milligrams
mg/L - milligrams per liter
mL - milliliters
mm - millimeters
mrem/yr - milliroentgen equivalents per man per year

NASQAN - National Stream Quality Accounting Network
NF - nanofiltration
ng/L - nanograms per liter
nm - nanometers
$NPSH_R$ - net positive suction head required
$NPSH_A$ - net positive suction head available
NTU - nephelometric turbidity units

oz - ounces
oz/gal - ounces per gallon

PAC - powdered activated carbon
pCi/L - picocuries per liter
PCT - physical chemical treatment
ppb - parts per billion
ppm - parts per million
ppt - parts per trillion
psig - pounds per square inch gauge

RO - reverse osmosis
rpm - rotations per minute

SHMP - sodium hexametaphosphate
SMCL - secondary maximum contaminant levels
SWDA - Safe Water Drinking Act (1974)

TAPPI - Technical Association of Pulp and Paper Industry
TDS - total dissolved solids
THM - trihalomethane
TNTC - too numerous to count
TTHM - total trihalomethanes

u.c. - uniformity coefficients
UF - ultrafiltration
USPHS - United States Public Health Service
UV - ultraviolet

VOS - volatile organic substances
vp - vapor pressure

WHO - World Health Organization

INDEX